INSECT POTPOURRI:

ADVENTURES IN ENTOMOLOGY

INSECT POTPOURRI:

Adventures in Entomology

Edited By

JEAN ADAMS

For

The Chesapeake Chapter of ARPE

The Sandhill Crane Press, Inc.
Gainesville, Florida

In house editor: Ross H. Arnett, Jr.
Production editor: Michael C. Thomas

ISBN: 1-877743-09-7

LIBRARY OF CONGRESS CATALOGING IN PUBLICATION DATA:

Insect potpourri : adventures in entomology / edited by Jean Adams for
   the Chesapeake Chapter of ARPE.
      p. cm.
   Includes bibliographical references.
   ISBN 1-877743-09-7 : 424.95
     1. Insect pests. 2. Insects. 3. Beneficial insects. 4. Insect
pests--Control. 5. Entomology. I. Adams, Jean R. (Jean Ruth)
II. American Registry of Professional Entomologists.
SB931.I572 1991   91-31846
595.7--dc20   CIP

The Sandhill Crane Press, Inc.
2406 NW 47th Terrace
Gainesville, FL 32606 USA

MANUFACTURED IN THE UNITED STATES OF AMERICA

PREFACE

The American Association of Economic Entomologists (AAEE) was organized over a century ago, in 1889, while the Entomological Society of America (ESA) was begun in 1906; in 1953 the AAEE merged with the ESA. Some years later, in 1969, the American Registry of Professional Entomologists (ARPE) of the ESA was established. An interesting account of this history is documented in the Centennial Edition of the ESA Bulletin, Volume 35(3): 1989.

In early 1988, the leadership of ARPE, the Governing Council, proposed the publication of a book conveying a generalized sense of entomology as a contribution to the 1989 Centennial Celebration of the ESA. While the project was unanimously endorsed by the Council, it was subsequently canceled because of the slow progress and the realization that the project could not be completed in time for the ESA celebration commemorating 100 years of "Entomology Serving Society."

The Chesapeake Chapter of ARPE, an independent group whose members are also long standing members of both the national ARPE and the ESA, undertook the completion of this project at the behest of the ARPE Governing Council. The Centennial Book Committee, established by the Chapter, contacted distinguished entomologists – some retired with years of experience as well as younger colleagues – requesting that they write some of their more interesting experiences in battling insect and related pests, accounts of "insects as friends," documentation of the role of insects and entomology in history, and in today's world, and other topics, for the general public.

Most chapters in the book are solicited manuscripts prepared especially for this publication; however, to round out the book, a few chapters are reprinted and are so acknowledged.

This book, "Insect Potpourri: Adventures in Entomology," is intended to be an informative, popularized version of entomology which will serve to make students and the general public more aware of and provide a better understanding of our profession. We hope all readers will gain some insight into man's dealing with the great diversity of the insect world and that those readers contemplating career choices will recognize the possibilities of a challenging and rewarding career in entomology.

v

The editors are grateful to the agents for permission to use the cartoons by Gary Larson. We also thank the companies listed below for their generous support.

ACKNOWLEDGEMENT

RHÔNE-POULENC INC.
Agrochemical Division
Research Triangle Park, North Carolina

S. C. JOHNSON & SONS, INC.
Racine, Wisconsin

NARAM CHEMICAL CO.
Wilimington, Delaware

Commemorative Book Committee, Chesapeake Chapter, ARPE:
   Jean R. Adams, Chair, *John J. Drea, John W. Kliewer, M. Joe Sloan, James H. Trosper,* and *William H. Wymer.*

TABLE OF CONTENTS

Chapter 6. Protecting Our Environment and the Challenge of Entomology as a Career, 221.

Chapter 7. Unusual Facts About Insects and Other Arthropods, 253

Names and Addresses of Contributors

Agnello, Arthur M., Department of Entomology, New York State Agriculture Experiment Station, Geneva, NY 14456 (see p. 192).

Baker, Edward W., Systematic Entomology Laboratory, U.S.D.A., A.R.S., Bldg. 003, BARC-E, Beltsville, MD 20705 (see p. 329).

Batra, Suzanne W. T., Beneficial Insects Laboratory, U.S.D.A., A.R.S., Beltsville, MD 20705 (see p. 15).

Bruce, W. G., Animal Disease Eradication Division, U.S.D.A., A.R.C., Washington, DC 20506 (see p. 266).

Carr, Richard V., Urban Biosystem, McLean, VA 22102 (see p. 128).

Coulson, Robert N., Department of Entomology, Texas A. & M. University, College Station, TX 77843 (see p. 109).

Cunningham, J. C., Forestry Canada, Forest Pest Management Institute, P. O. Box 490, Sault Ste. Marie, ON P6A 5MT, Canada (see p. 112).

Davich, Theodore B., USDA, ARS (Retired), 110 Fox Hill Drive, Starkville, MS 39759 (see p. 186).

DeFoliart, Gene R., Department of Entomology, University of Wisconsin, Madison, WI 53706 (see p. 44).

Drea, John J., Beneficial Insects Laboratory, U.S.D.A., A.R.S., Beltsville, MD 20705 (see p. 18, 227).

Dutky, Samson R., U.S.D.A., A.R.S. (Retired), 14106 New Hampshire Ave., Silver Spring, MD 20904 (see p. 236).

Frazier, Norman W., University of California (Retired), 5104 Hamel St., Davis, CA 95616 (see p. 209).

Gerberg, Eugene J., 5819 57th Way, Gainesville, FL 32606 (see p. 243).

Gordon, Robert D., Systematic Entomology Laboratory, Smithsonian Institution, NHB 168, Washington, DC 20560 (seep. 24).

Gorham, J. Richard, Food and Drug Administration, 200 C St., S.W., Washington, DC 20204 (see p. 124).

Greenberg, Bernard, Department of Biological Sciences, University of Illinois, Chicago, IL 60680 (see p. 97).

Grissell, Eric, Systematic Entomology Laboratory, NHB 168, Smithsonian Institution, Washington, DC 20560 (see p. 248).

Hamel, Dennis R., deceased (see p. 259).

Hanski, Ilkka, Department of Zoology, University of Helsinki, SF-00100 Helsinki, Finland (see p. 278).

Harris, Marvin K., Department of Entomology, Texas A. & M. University, College Station, TX 77843 (see p. 240).

Hardy, Tad N., Department of Entomology, Louisiana State University, Baton Rouge, LA 70821 (see p. 307).

Hermann, Henry R., Department of Entomology, University of Georgia, Athens, GA 30602 (see p. 2).

Hollis, William L., Pesticide Assessment Laboratory, U.S.D.A., A.R.C, Beltsville, MD 20705 (see p. 170).

Kliewer, John W., Environmental Protection Agency (Retired), 9805 Meadow Knoll Ct., Vienna, VA 22181 (see p. 222).

Knipling, E. F., U.S.D.A. A.R.S. (Retired), 2623 Military Road, Arlington, VA 22207 (see p. 84).

Koehler, Philip G., Institute of Food and Agriculture, University of Florida, Gainesville, FL 32611 (see p. 147).

Kritsky, Gene R., Department of Biology, College of Mount St. Joseph, Mount St. Joseph, OH 45051 (see p. 40).

Larson, Larry L., Dowelanco Research Laboratories, Greenfiled, IN 46140 (see p. 162).

Leung, Harry, Department of Zoology, University of Western Ontario, London, ON N6A 5B7, Canada (see p. 254).

Linduska, James J., University of Maryland, Vegetable Research Farm, Quantico Road, Salisbury, MD 21801 (see p. 175).

Linduska, Joseph P., University of Maryland, Vegetable Research Farm, Quantico Road, Salisbury, MD 21801 (see p. 175).

Locke, Marius, Department of Zoology, University of Western Ontario, London, ON N6A 5B7, Canada (see p. 254).

Locke, Michael, Department of Zoology, University of Western Ontario, London, ON N6A 5B7, Canada (see p. 254).

Lofgren, Clifford S., U.S.D.A., A.R.S. (Retired), 1321 N.W. 31st Drive, Gainesville, FL 32601 (see p. 141).

Martin, Dial F., U.S.D.A., A.R.S. (Retired), 3600 Sierra Drive, Bryan, TX 77802 (see p. 182).

Matthews, Janice, College of Veterinary Medicine, University of Georgia, Athens, GA 30602 (see p. 321).

Mauldin, Joe K., U.S.D.A. Forest Service, Southern Forest Experiment Station, P. O. Box 2008, GMF, Gulfport, MS 39505 (see p. 138).

Mertins, James W., U.S.D.A., A.P.H.I.S., N.V.S.L., P.L., P. O. Box 844, Ames, IA 50010 (see p. 281).

Metcalf, Robert L., Department of Entomology, University of Illinois, Urbana-Champaign, IL 61801 (see p. 203).

Morse, Roger A., Department of Entomology, Cornell University, Ithaca, NY 14853 (see p. 6, 151).

Mulrennan, John Andrew, Jr., Department of Health and Rehabilitative Services, P. O. Box 210, Jacksonville, FL 32231 (see p. 75).

Nordlund, Donald A., U.S.D.A., A.R.S., S.A.R.L., B.C.P.R.U., 2413 East Highway 83, Weslaco, TX 78596 (see p. 213).

Patterson, Richard S., Insects Affecting Man and Animals Laboratory, U.S.D.A., A.R.S., 1600 S.W. 23rd Drive, Gainesville, FL 32604 (see p. 147).

Pratt, Harry D., 879 Glen Arden Way, N.E., Atlanta, GA 30306 (see p. 56).

Rambo, George W., Director, National Pest Control Association, Dunn Loring, VA 22027 (see p. 134).

Raun, Earle S., 3036 Prairie Road, Lincoln, NE 68506 (see p. 73).

Schreck, Carl E., U.S.D.A., A.R.S., Insects Affecting Man and Animals Research Laboratory, P. O. Box 14565, Gainesville, FL 32604 (see p. 79).

Shimanuki, H., Beneficial Insects Laboratory, U.S.D.A., A.R.C., Beltsville, MD 20705 (see p. 34).

Showers, William B., U.S.D.A., A.R.S., Corn Insect Research Laboratory, Department of Entomology, Iowa State University, Ames, IA 50011 (see p. 199).

Shultz, Harvey A., C.O.D.E., 143 Northern Division, Navfacengcom Bldg., 77 Naval Base, Philadelphia, PA 19112 (see p. 61).

Sosa, Omelio, Jr., Sugarcane Field Station, U.S.D.A., A.R.S., Canal Point, FL 33438 (see p. 50).

Sutherland, Douglas W. S., Chairman, National Insect Subcommittee, Entomological Society of America, 9301 Annapolis Rd., Lanham, MD 20706-3115 (see p. 28).

Tompkins, George J., U. S. Environmental Protection Agency, 401 M. Street, S.W., Washington, DC 20460 (see p. 216).

Trosper, James H., Defense Pest Management Information Analysis Center, Walter Reed Army Medical Center, Washington, DC 20307 (see p. 293).

Webb, Ralph E. Insect Biocontrol Laboratory, U.S.D.A., A.R.S., Beltsville, MD 20705 (see p. 102).

Chapter 1

Insects as Friends

THE FAR SIDE   By GARY LARSON

© 1991 Universal Press Syndicate

"Got him, Byron! It's something in *the Vespula* genus, all right — and oooooweeeee does he look mad!"

To Bee or Not to Bee Social

Henry R. Hermann

Sociality in insects, as well as in animals in general, is actively being researched by many of today's ethologists and theoretical biologists, and while there are certain concepts involving sociality that are currently well understood, there are many questions that must be answered. In short, the more we find out about these fascinating animals, the more interesting they seem to be.

The state of being social has been variously defined through the years. Based on its Latin root, the term "social" comes from "socialis," meaning merely to associate. However, it has been emphasized on numerous occasions that many insects associate with one another, but other than the fact that the insects are together, their behavior actually provides scant reason to call them truly social. Such associations, generally referred to as aggregations, may form for mating purposes, group protection, shelter and many other reasons. However, scientists have found that insects generally regarded as social have some other outstanding qualities about them that sets them apart from most other insects.

It appears that there are a number of advantages for being social, such as cooperation between individuals to increase a species' productivity through group defense, food gathering, nest building, and care of the young. There are also certain obvious disadvantages, such as more intense predation, parasitism and parasitoidism. This raises many questions about how and why insects and other animals developed a social organization.

Within the last 50-60 years, interest in social behavior has increased, and new definitions have come and gone. For example, with the publication of a book in 1962 called "Animal Dispersion in Relation to Social Behavior," a society was broadly defined by the author, V. C. Wynn-Edwards, as "an organization capable of providing conventional competition. Conventional competition is competition for conventional rewards such as food, mates, territories, shelter and status." If we use this definition as our guide to understanding sociality, we find that most, if not all, animals appear to be social in one way or another and that the type and degree of competition defines the level of sociality that is occupied by a particular species.

In 1971, E. O. Wilson more precisely defined the maximum degree of sociality in insects as eusociality in his book called "The Insect Societies." According to his definition, eusocial insects possess three outstanding features:

1) eusocial insects care for their young; in fact, the young absolutely require care because they are totally dependent on the adults for all of their external needs;

2) there is an overlap in generations, that is, the parents and offspring live together for some period of time and interact with one another;

3) there is a reproductive division of labor, meaning that there is an egg-laying female and other females that function in other ways.

When examining the class Insecta, we realize that only a small percentage of insects are actually eusocial and that eusociality exists only in two of the 35 or so recognized insect orders. All of the termites (order Isoptera) and some of the order Hymenoptera (ants, bees, and wasps) are eusocial: all ants, honey bees, stingless bees, bumble bees, some members of a less well known group referred to as the allodapine bees, some of the sweat bees, at least one wasp species in the family Sphecidae, and the paper wasps.

A vast number of bees and wasps are not eusocial, but some of these insects and certain members of other insect orders may possess one or two of the characteristics used to define eusociality. For instance, many non-eusocial insects both inside and outside of the orders Isoptera and Hymenoptera take care of their young.

In 1975, E. O. Wilson redefined sociality, but this time in a broader way in his book, "Sociobiology, A New Synthesis." However, he was at this time concerned with all animal groups rather than just insects. According to this treatment, a society "is a group of individuals belonging to the same species and organized in a cooperative manner." While this definition is currently useful, it remains somewhat vague because many animals can form aggregations in which they cooperate for one reason or another. Certain caterpillars, for example, can aggregate on their host plant and cooperate in a group defensive strategy.

Societies generally form colonies in which individuals remain together, care for their offspring and cooperate to do certain tasks. However, there is some lack of cooperation during certain stages of colony life, such as during the early development of a colony when females of certain eusocial species compete for a position of dominance. There is a caste system, that is, there are females that have different functional roles (feeding their young, building their nest, foraging for food, defending the colony).

The field of sociality has become quite active since the 1964 and 1972 papers of W. D. Hamilton brought to light the possibility of a genetical basis for the rise and maintenance of sociality. Since those publications appeared, numerous individuals have studied sociality from various views, including degrees of relatedness, inclusive fitness, parental manipulation, and others.

While it appears obvious that closely related individuals are more prone to be altruistic toward each other than are those individuals that are distantly related, the idea of relatedness becomes confusing in insects that have multiple matings. A female that is multiply mated by different males will bear young that are not as close genetically to one another as those that are produced from a female that mated with only one male.

Likewise, degrees of relatedness are confused by the presence of multiple queens (polygyny) and egg-laying workers (gamergates). Some workers are known to be involved with thelytokous (female producing) and arrhenotokous (male producing) parthenogenesis. There are also cases of ergatoid (worker-like) individuals that are actually mated with males in their own colony and thus are capable of producing offspring much like those of the queen.

The production of males by workers in the order Hymenoptera is immediately understandable since reproduction is based on haplodiploidy (males are haploid while females are diploid). Dominant workers, in certain cases, may even produce most or all of the male offspring in their colony. However, female production by workers requires either that the germinal cells fail to separate during their first meiotic division or that there is a fusion between two cells prior to egg deposition.

These are but a few of the numerous problems being examined today by researchers of eusocial insects. Add to this the study of dominance hierarchies (peck orders), defensive tactics, social parasitism, the development of models for social insect strategies, the examination of relationships between taxa by using electrophoresis, reclassification based on new criteria, and the construction of cladograms to represent social insect phylogenies, we find that social insect biology is currently a very active and exciting field in which to work. Much is still to be done with social insects in the fields of taxonomy, behavior, genetics, morphology, physiology, phylogeny, and other areas before we can tie together many of the concepts that are currently being hypothesized.

Suggested further reading

Brian, M. V. 1983. Social Insects, Ecology and Behavioral Biology. Chapman and Hall, New York.

Hermann, H. R. 1979-1982. Social Insects, Vols. 1-4. Academic Press, New York.

Oster, G. F., & E. O. Wilson. 1978. Caste and Ecology in the Social Insects. Princeton Univ. Press, Princeton, NJ.

Wilson, E. O. 1971. The Insect Societies. Harvard Univ. Press, Cambridge, Mass.

Wilson, E. O. 1975. Sociobiology, A New Synthesis. Harvard Univ. Press, Cambridge, Mass.

Wynn-Edwards, V. C. 1962. Animal Dispersion in Relation to Social Behavior. Hafner Pub. Co., New York.

Hobby and Commercial Beekeeping In the United States

Roger A. Morse

The American agricultural scene is changing rapidly. While 30 percent of our population is involved in food production, harvesting, processing, storage, distribution, and sale, fewer than two percent of our people are concerned with production and harvesting. The result of this intense specialization is that no where on earth is food more delicious, nutritious, abundant, diverse, safer, and cheaper than in the United States. Part of this agricultural revolution involves the use of honey bees for pollination. It is important too, to understand that this revolution is not slowing but if anything is becoming more intense.

In many areas in agriculture the family farm is being replaced by corporate farming. This is less true in beekeeping where a precise management scheme is demanded. While the bulk of the bees rented for pollination are cared for by the larger, commercial beekeepers, a number of hobbyists have found it profitable to rent their few colonies to a nearby fruit, seed, or nut producer who will often transport the bees in addition to paying a fee for their use.

An example of this change in agriculture is the number of colonies of honey bees taken into the state of Maine each year for blueberry and apple pollination, most for use on blueberries. There were 8,000 colonies rented to Maine growers for pollination in 1981. In 1989 that number was just a few over 30,000. Most of these colonies were carried into the state on trucks from Florida. After the blueberries are finished flowering in late May the colonies are carried south to Massachusetts where they are used for the pollination of cranberries that grow in bogs, mostly in the Cape Cod region.

A few over one million colonies of honey bees are rented for commercial pollination each year in the U.S.; most of these are used an two crops in the spring of the year and thus the beekeeper receives two rental fees per colony. The greatest number of colonies, about 70 percent of the total, are rented in California. In that state 650,000 colonies are used to pollinate a little over 400,000 acres of almonds alone, but bees are needed to pollinate other crops as well. The number of colonies of honey bees being rented today for pollination is estimated to be four times the number rented ten years ago and ten times the number used twenty years ago. The crops for which bees are rented are listed in Table 1.

Table 1. Estimated number of colonies of honey bees being rented for pollination annually in the U.S.

Crop	Colonies Rented
almonds	650,000
apples	250,000
melons	250,000
alfalfa seed	220,000
plums/prunes	125,000
avocadoes	100,000
blueberries	75,000
cherries	70,000
vegetable seeds	50,000
pears	50,000
cucumbers	40,000
cranberries	30,000
kiwifruit	15,000
all others	50,000
Total	2,035,000

There are over 200,000 people who own a hive or more of honey bees in this country. Only about one percent, about 2,000 people, are beekeepers who make a full-time living keeping bees. These beekeepers usually own 600 to 2,000 colonies each but there are many who own and operate more, some as many as 20,000 to 40,000 colonies. Many of the these commercial beekeepers are migratory and carry their bees south in the fall for winter and then move them back north for pollination or honey production. Thus, both the beekeepers and the bees enjoy warm weather throughout the year. A New York State beekeeper, for example, may spend the late spring and summer months first pollinating apples and then producing honey from clover and alfalfa. In the fall he may move his family and bees to Florida where a crop of orange honey may be secured in the early spring before they all return north about mid-April.

A number of beekeepers are sideliners who own 50 to a few hundred colonies, that again may be rented for pollination, used in honey production, or both. Most beekeepers, however, are hobbyists who own one to ten hives that they keep because they enjoy an outdoor hobby and the challenge of managing colonies for honey production. Some of these beekeepers may sell the honey their bees produce but most use it on their own table, in cooking, to make honey wine or some other product, and to give to friends. There is always a ready market for beeswax though many beekeepers enjoy making beeswax candles for sale, for themselves, or as gifts for friends.

Another small group of beekeepers are biologists who have discovered that the honey bee is an ideal experimental animal whose nest may be easily moved and manipulated. Honey bees have the most advanced system of communication that has been found among lower animals; while we do not understand all aspects of this system we have made major advances in our knowledge about colony organization, division of labor, defense, foraging strategy, and reproduction. All of the activities in a colony may be seen and studied using glass-walled observation hives that may be kept in a laboratory or living room.

The most important consideration in successful beekeeping is food for the bees. One must keep their bees where there are natural sources of pollen and nectar for these are their sole foods. Honey is the honey bees' source of carbohydrate, and pollen supplies the protein and small amount of fat that is needed. Beekeepers do not plant crops for their bees, rather, they move to areas where honey and pollen-producing plants abound. And, of course, it is necessary that the bees have reserves of food or that fresh nectar and pollen, especially pollen, is available throughout the active season whenever bees can fly. A single colony of bees may be kept almost anywhere though some forested areas and innercities may have little to offer in the way of food. A small number of colonies are kept on Manhattan, the island borough of New York City, but obviously the number is limited because there are so few flowers in the city for the bees to feed on. In the U.S. as a whole, about half of the nectar that honey bees collect and make into honey, comes from weed plants such as goldenrod, sumac, thistle, sage, gallberry, and trees that line our streets or grow in forests such as some maples, oaks, basswood, and sourwood. The other half of the honey that is produced comes from cultivated plants such as citrus, cotton, rape, sunflowers, or clovers and alfalfa grown for seed or hay.

Honey is a unique food that has played an important role in our history. Sugar cane was first found on the South Pacific islands by the Chinese who built the first sugar mill over 2,000 years ago. At that time, however, there was little exchange of people or goods between East and West and it was not until about 600 years ago that sugar cane was brought into the Mediterranean area and Europe. Prior to that time honey was early man's chief sweet in that part of the world. It has been only in the past 50 to 75 years that most people have been able to obtain all of the sugar they wanted and that their sweet tooths have been satiated. Thus, until recently, honey has played an important role in this regard. Even today it is much sought after by many people.

Honey bees make honey from nectar. Most nectars contain water and the sugar sucrose, the same sugar that is found in cane or table sugar. Nectar also contains plant pigments, nutrients, and other components so that each nectar is different just as flower colors, designs, and shapes are different. Once they have found a good sourse of food, bees use these different markers to identify and go from a flower of one kind to another of the same kind so as to maximize the quantity and quality of the food they may gather. To make honey, bees reduce the moisture content to between 16 and 19 percent. At this low sugar concentration honey has a high osmotic pressure and microbes cannot survive in honey because they lose water through their cell walls and die. Additionally, bees add an enzyme, invertase, that splits the sugar sucrose into two sugars each with half of the number of carbon atoms found in sucrose. These two sugars are glucose which is the same sugar found in corn, and fructose. Fructose is the sweetest of the common sugars and is found in many fruits. A second enzyme that bees add is glucose oxidase. This attacks a very small portion of the glucose and converts it into hydrogen peroxide and gluconic acid. The glucose oxidase system works only when the honey is unripe or is diluted by the bees as it is when they feed their larvae. Hydrogen peroxide, which is a strong antibacterial agent, is an unstable compound and is not found in ripe honey. However, the gluconic acid, which is made as a result of the enzymatic action, makes the honey an acid medium, as acid as wine or vinegar. The fact that honey is so acid, even through the total amount of acid present is small, serves to protect the honey against microbes that find it an inhospitable medium for growth for that reason too.

A few people have complained that honey is too sweet. This is for two reasons: one is that the fructose, which is, as mentioned, the sweetest of the common sugars. The second reason is the acid nature of honey, which together with the sugar, stimulates the taste buds in much the same way as does an acid cola drink or a Chinese sweet and sour, pickle sauce. It takes much less honey to satisfy one's carving for sweet than it does glucose, which is the least sweet of the common sugars. In fact, candy manufacturers use glucose in their products so that people can eat more candy at one time.

A major problem for beekeepers is the control of the pests, predators, and diseases that might ravage a colony of honey bees. A colony may have 50 to 200 pounds of honey and a much smaller quantity of pollen stored at any one time. This represents a great warehouse or food reservoir for hungry animals and microbes. While bears are often thought of as major predators, and they are in some areas, it is the viruses, bacteria, fungi, and mites that are a much greater problem. In the past 18 years three organisms that cause serious honey bee diseases have entered the country from abroad.

The first of these newly introduced diseases is caused by a fungus that kills bees in the larval stage. The disease is called chalkbrood because of the bright, white appearance of larvae when they are first infected and killed. Chalkbrood was first found in California in 1972 and within two years was found everywhere in the country. Chalkbrood, and the other diseases of honey bees, have been spread rapidly because of the extensive movement of bees by migratory beekeepers as well as those who grow queen honey bees and worker bees that are sold by the pound wherever they are needed. When chalkbrood was first found in the U.S.A. it was responsible for the deaths of thousands of colonies. Some races of honey bees are far more susceptible to chalkbrood than others. No treatment for the disease is known but beekeepers are now selecting among their colonies for bees that show some degree of natural resistance and the problem is much less serious than it once was thought, however, it is still of concern.

A mite, an eight-legged creature that can be seen only with a microscope, was found in Mexico in about 1980 and in the vicinity of Brownsville, the southern most city in Texas, on July 3, 1984. This mite is called the tracheal mite since it infests the breathing tubes of the honey bee both blocking them and also allowing bacteria to enter through the puncture holes the mite make in these tubes. This mite has long been a scourge in Europe, especially Great Britain, where it apparently originated. Between about 1900 and 1920 over 90 percent of the honey bee colonies in that country died because of this mite. However, the colonies that survived have a certain degree of natural resistance. The honey bees in North America had never been exposed to this mite prior to 1984 and in the past few years a large number of colonies have died from mite infestations. In the summer of 1989, 26 queen honey bees from Great Britain were brought into this country to Cornell University and are currently being used to grow stock that we hope will be resistant to tracheal mites. It has been discovered that menthol, a natural product extracted from the leaves of mint, which is used to flavor gum and baked goods, is toxic to the mites. However, for the fumes of menthol to be effective the chemical must be applied when the weather is warm. The fall is the ideal time to treat colonies; however, it is often too cool for the menthol to evaporate and cause the death of the mites in the northern states.

A third new problem for beekeepers in North America is the varroa mite whose seriousness in European honey bees was discovered only in the early 1960s in Asia. In 1971 some Japanese beekeepers gave a gift of live bees to beekeepers in Paraguay. At that time the problems the mite might cause were little known, and, in fact, it was not known that the bees that were sent from Japan were infested. The varroa mite is about the size of a pinhead. The mites attack larval, pupa, and adult honey bees and may kill or maim them. Those honey bees that survive a mite attack usually live for a much shorter period of time. By 1973 the varroa mites had been accidently spread by beekeepers over much of South America and were moving northward rapidly. They were found in Wisconsin on September 25, 1987 but it was soon found that they were widespread in the U.S. and there was no action that could be taken to prevent further spread. A chemical treatment to kill these mites has been developed. Beekeepers, who are proud of the fact that they use so few chemicals in their industry and that honey is a pure, natural product, are reluctant to use chemicals for control, though in some areas in the country they have been forced to do so. The varroa mites are also widespread in Europe where they were accidently introduced much earlier; we are now benefitting from the research that has been taking place there for many years. The experience in Europe shows that without a chemical treatment all of the honey bees will die.

The native host of the varroa mites is an Asian honey bee that is about half the size of our own honey bee. The Asian bees do not suffer seriously from the varroa mites and keep the mite numbers at a low level by grooming and physically picking the mites up with their mandibles, biting them to break their hard, outer shell, and then carrying them outside of the hive and unceremoniously dumping them on the ground where the mites die. This simple grooming behavior is ample to protect the Asian honey bees against these mites. We are now making observations on colonies in this country that might show signs of this good housekeeping. Interestingly, the Africanized honey bees in South and Central America, that are infested with these mites, do not suffer seriously from them. In fact, it has been suggested that the Africanized bees may be of benefit to North American beekeepers because of their natural resistance to the varroa mite. We are not certain how the Africanized bees control the varroa mites; we know only that we find them in every colony where we look for them in Brazil. Beekeepers there do not consider the mites a problem and do not treat their colonies in any way for them. In Africanized honey bee colonies the populations of varroa mites always remain low. Africanized honey bees, while they sting excessively and are not easy to manage, have certain traits that may make them of value in the U.S.; one of these is their apparent natural resistance to varroa mites and certain other honey bee diseases.

Despite the rapidly increasing need and use of honey bees for pollination, and the problems brought about by honey bee pests, predators, and diseases, honey production is still the major focus for the majority of beekeepers in the U.S. We produce about 250,000,000 pounds of honey each year, an average of only a little over one pound per person. Honey ranges in color from water white to coal black; some honeys have a reddish hue and others are greenish though most are a typical yellow or amber in color. One honey in North Carolina, made from plants that grow on soils with a high aluminum content, is sometimes blue in color. The lighter colored honeys are usually milder in flavor and are used, for the most part, as table honey to be eaten by itself, on muffins or cakes, in coffee, or as a topping on ice cream. About half of the honey produced in the country is used as table honey while the other half is used by the bakery industry. Honey is especially popular in graham crackers though more recently it has appeared as an ingredient in cereals, specialty cakes, and cookies.

Beekeeping is especially attractive as a hobby because it does not require a large investment in land, machinery, buildings, and processing equipment. Other than the bees themselves, an extractor, a centrifugal force machine that is used to throw the liquid honey from the combs in which it has been stored by the bees, may be the most expensive item one needs to buy. However, many hobby beekeepers share extractors with one another, while some beekeepers will extract the honey for others for a small fee. If one invests in hobby beekeeping carefully it is possible to obtain a ten to twenty five percent return on one's investment almost every year. That is not true of very many hobbies. The most important consideration has already been mentioned and that is having an abundance of nectar producing plants in the vicinity so that the bees may gather a surplus over their daily needs and the food they need to survive the winter. Many people have entered hobby beekeeping and have slowly increased their colony numbers until they have reached the point they could make beekeeping a full-time vocation.

The best way to start in beekeeping is to buy one or more second hand colonies of bees. Hopefully, the beekeeper from whom the bees are purchased will be willing to give advice and some help to the a beginner starting his or her hobby. One of the chief advantages of buying a second hand colony is that it should already be a mature colony and ready to produce it's owner a surplus of honey the first year. As a beginner it is really best to buy three to five colonies so that one can become familiar with how colonies may vary; when one has only one colony it is difficult to know if the population is normal for the season.

A second way to start in beekeeping is to buy two or three pounds of bees and a queen, called a package, from a southern or California beekeeper. If package bees are purchased they should be brought north and put into a hive at the time the first pussy willow blooms. In states such as New York, Michigan, and Wisconsin this usually means about April 15. The first year a package of bees needs time to build its population and as a result it is rarely able to produce a surplus for the beekeeper.

There are three other ways one may obtain bees. One is to contact the local fire and police departments who are always looking for beekeepers willing to collect stray swarms. One may also remove bees from a building or a tree; doing so, however, is very hard work and often not successful, as it is difficult to find the queen in a natural nest where the comb must be cut and removed piece by piece, without hurting or damaging her. A last method of starting in beekeeping is to use bait hives, boxes that are made and hung in trees in such a way as to be attractive to a stray swarm. Old, second-hand and worn hive boxes, called supers, make excellent bait hives. Bait hives must have tight, light proof covers and be hung in trees at a height of five or so yards where they are visible but shaded. Honey bees have a remarkable ability to measure the quality of a bait hive and will reject those that are not suitable.

Many sources of information on bees and beekeeping are available. Each state has a state agricultural college and most of these offer a course in beekeeping as well as one or more bulletins and leaflets on beekeeping within their boundaries. The U.S. Department of Agriculture also has several publications. Most states have statewide beekeeper's organizations. Perhaps most important, however, are the numerous local organizations that usually work with or through the county extension service. There is a county agricultural agent or specialist in every county in the U.S.A. and these people usually have access to a wealth of information on bees and beekeeping. There are two national trade journals and one national beekeeping newspaper. In addition to carrying articles about beekeeping these will also list upcoming meetings and events. Several states and and local beekeepers organizations offer short courses in beekeeping, There are two national beekeeper's organization: The American Beekeeping Federation whose membership is open to all beekeepers, honey packers, bee supply dealers, and others interested in the industry, and The American Honey Producers whose members are commercial honey producers and pollinators. On a regional basis there is also an Eastern Apicultural Society that serves the northeastern states and easternmost Canadian provinces, a Western Apicultural Society that serves the westernmost states and Canadian provinces, and the Southern States Beekeeper's Federation. All of these groups hold meetings at least once a year and most will have honey shows, commercial exhibits, and often will offer a beekeeping short course. Many beekeepers groups hold their meetings in members' apiaries in the summer where colonies will be opened and one may gain hands-on experience. There is also a great abundance of books on beekeeping, some regional, and some that cover beekeeping in the whole country. Again, the county agricultural extension specialist is the best source of information.

Bees and Pollination

Suzanne W. T. Batra

Everyone knows the popular honey bee that stores sweet honey in waxen combs in hives. A few people know that honey bees are not native to North America, but came here from Europe with the settlers. The African or "killer" bee is a tropical variety of honey bee that was brought by a scientist to Brazil from Africa in 1956. These honey bees have been breeding with European honey bees, multiplying rapidly, and spreading northward through Mexico. Beekeepers are worried that many of them may have to go out of business or raise the cost of renting hives to growers for pollination, due to the difficulty of handling African honey bees, legal problems, and mite parasites of honey bees that have recently arrived here.

However, honey bees have numerous relatives that can also be used for the pollination that is needed to produce many crops. There are about 21,000 kinds, or species, of bees (Apoidea) in the world, and 3,500 of these live in North America. Most wild bee species are solitary, which means that each female bee makes her own nest and a few brood cells, and stocks each cell with nectar, pollen, and an egg. These bees have no queens and workers, as well as no honey, beeswax, or beehives. Instead, wild bees make their small nests in holes in wood, in plant stems, or underground. Because all bees visit flowers for their nectar and pollen diet, all are valuable pollinators, and thus they improve the yield of both wild and cultivated plants.

You may already know the fuzzy, buzzy, bumble bees; the big, black, hole-drilling carpenter bees; and the pesky little sweat bees. Like other wildlife, bees need safe places to build nests and raise their young. Populations may dramatically decline when fallow land is cultivated or when pesticides are used. When this happens, yields of seeds, fruits, and vegetables will fall, unless hives of honey bees or other pollinating bees are put in the fields to replace the original native wild bees. Although most research concerns honey bees, enough has been done by now so that we know that there are many kinds of bees that can be semi-domesticated and used for pollination service.

Hornfaced bees (*Osmia cornifrons*) are solitary bees that have been used in Japan as orchard pollinators since the 1930's. They are quite unlike honey bees in behavior, and thus they must be managed differently. Most of the year, they are dormant, and the adult bees forage and pollinate only during spring. Only one generation of bees is produced each year. After mating in early spring, each female constructs rows of cells made of mud inside cardboard tubes or reeds supplied by the beekeeper. Each mud cell is provisioned by the female with a mixture of nectar and pollen, on which she lays an egg. Each cell is sealed with mud, and the larva feeds and develops without further attention from its mother.

These bees are valued because of their superior ability to pollinate. Few of them are needed per acre of orchard because they are so efficient. For example, an individual hornfaced bee can set an average of 2,450 apples per day, but an individual honey bee sets only 30 per day. This difference occurs because hornfaced bees work faster than honey bees, stay within the orchard, contact stigmas consistently, and prefer orchard flowers to those of weeds or other crops. They are commercially used in Japan to pollinate apples, cherries, plums, peaches, and pears, and they have been imported into the U.S. and Canada. A close relative, the native North American blue orchard bee, *Osmia lignaria*, has also recently been managed for almond pollination in the western U.S.

Alfalfa is one of our major livestock forage crops, and the supply of seed is a $100-million-per-year industry. The flower of alfalfa has a little spring in it that is triggered when a bee pushes her head into the flower. The stamens pop up and rap the bee on the head, dusting her with pollen. Honeybees apparently do not like being rapped on the head, and since they are fairly intelligent insects, they learn to poke their long tongues into the sides of the flowers, avoiding the spring. Thus they take nectar but accomplish no pollination. In the northwestern U.S. and Canada, two species of solitary bees, the alfalfa-leafcutter bee and the alkali bee, are intensively managed to pollinate alfalfa since the 1950's. These persistent insects do not seem to mind being rapped by the flowers, and they collect the pollen for their nest cells as it is showered on them. The small, gray, originally European, alfalfa-leafcutter bee quickly visits many flowers in succession and is a highly efficient pollinator. It nests in groups. Farmers make suitable holes in "bee boards," which they put in shedlike shelters in the alfalfa fields. The brood cells of this bee are built out of bits of alfalfa leaf.

The alkali bee is a blue-banded species of sweat bee native to the U.S.A. It likes to nest in alkaline flats, where the subsoil stays fairly moist. Growers who accommodate this bee construct "bee beds," which consist of plastic-lined trenches in the alfalfa fields. The trenches are filled with specially prepared soil that is maintained at just the right level of moisture to attract and hold dense aggregations of alkali bees. There may be as many as 2,100 nests per square meter.

It is good to conserve and encourage wild native bees. They can contribute significantly to pollination if they are protected from insecticides, and if zones of uncultivated land with nesting sites are retained next to crops. Fruit trees that grow along the edges of orchards often host swarms of wild pollinators of diverse sizes, shapes, and colors. Most people don't think that they are bees, because many resemble wasps or flies rather than honey bees. You can know that they are bees if they collect pollen. It is fun to go out on a sunny spring day, just to take a look at them working hard for us.

Suggested Further Reading

Batra, S. W. T. 1984. Solitary Bees. Scientific American 250(2): 120-127.

Insects in White Hats - The Good Guys of Agriculture

J.J. Drea and R.M. Hendrickson, Jr.

She treated the squirming little bug like it was a delicate kabob. She held onto it, twisted it around and around, as if to locate the most tender part. Then she stabbed it with her needle-like stinger and deposited an egg deep inside the little creature. The tiny bug dropped to the ground, and lay paralyzed for the longest time. Its days were numbered. Who was the assassin that struck such a foul blow to this tiny insect?

The perpetrator of this deed was a tiny wasp from Turkey, scarcely larger than the plant bug in which it planted the seed of death. But this wasp was no unwelcome visitor to our shores. It had been purposely imported from Asia into the United States by American scientists as a weapon in the war against insect pests.

Its victim was a species of insect, called a Lygus bug or tarnished plant bug, that causes millions of dollars in damage and loss to U.S.A. crops each year. The pest bug sucks the sap from all types of plants, from alfalfa to peaches. Damage by the insect usually is not evident until the buds or fruit, fed on weeks before, fail to develop, or are deformed and not marketable. Feeding by lygus bugs can totally destroy a seed crop.

The killer-parasite of this scene lived and reproduced at the expense of the bugs it captured. The egg it had inserted soon hatched into a tiny maggot-like individual that fed inside the body of the bug. Eventually, the tiny immature parasite ate all the non-essential parts of its host. Then it turned to the vital organs, finished them off, bored a hole through the side of the dying bug, squeezed through the opening, dropped to the ground, and formed a cocoon in the soil. Eventually, a fully-formed wasp emerged from the cocoon and the murderous cycle began again.

Battles more vicious than that between the wasp and the lygus bug contantly rage around us. The most deadly foe of an insect usually is another insect. Thousands and thousands of different types of insects constantly pursue other insects to lay eggs on and in them, paralyze and store them for food for their offspring, or to feed directly on them. These conflicts are unending, pityless, and without any truce. Often the outcome of these battles are of vital importance to U.S.A. farmers who generally are completely unaware that these conflicts occur.

However, not all of this intense life and death struggle has gone unnoticed by man or without human intervention. The Chinese of antiquity observed that some ants could be used to battle sapsucking insects feeding on their citrus trees. These clever people even constructed bridges between trees to permit the ants to outflank and capture their insect prey.

Observers began to realise that some groups of insects — the ones in "white hats" — habitually fed on other insects, the villains in "black hats." Furthermore, when this happened the plants involved, especially food crops, often were better off as a result. The "ladybug, ladybug fly away home" plea of the ageless rhyme refers to the beneficial lady beetles that were known to feed on aphids infesting hopyards in Europe.

Not until 100 years ago did this battle between the beneficial insects and the villians take on special significance, especially in the United States. After the American Civil War, a foreign insect pest invaded the citrus groves of California. This unwanted immigrant resembled a tiny fluff of cotton and was called the cottonycushion scale. This pest was in paradise — there were no natural enemies to worry about and thousands of acres of trees available for food.

This voracious pest, apparently from "Down-under," inserted its needle-like beak into all parts of a citrus tree and fed on its juices. The fruit was destroyed and often the trees were killed. In no time, citrus production declined to alarming levels and the California citrus industry itself was in jeopardy.

Along came an insect in a white hat to the rescue. A tiny lady beetle, known as the Vedalia Beetle, was found and collected in Australia and New Zealand and brought to the United States by an American entomologist. This little predator craved the cottonycushion scale and gorged on the pest. The female laid her eggs in the body of the insect and the offspring grew to adulthood feeding on a diet of the scale.

When released in the groves of California the impact of the ladybeetle was tremendous. The Vedalia fed on the scale and multiplied by the thousands. The villain was decimated and the citrus industry was snatched from the brink of destruction. Joy returned to the orange growers of California. To this day the little beetle is still there waiting to pounce on any cottonycushion scale with the audacity to appear in the neighborhood.

The Vedalia affair encouraged others to try to control insect pests with other insects. The field of biological control came into being as a recognized science. Unfortunately, this approach did not turn out to be a panacea for all our pest problems. There were many failures in the succeeding years. Nevertheless, efforts to control unwanted foreign pests with natural enemies did continue. As a result many exotic insects no longer pose a threat to American agriculture.

Then, during the '50s, came the era of potent chemical pesticides. Not only did these chemicals kill insects, they just about killed biological control. Who needed good bugs when a good spray got them all? The eventual outcome is known to us all. The insects fought back with resistance. Chemical warfare suffered a major defeat.

The public and its elected officials, agency administrators, and scientists began to demand and search for other ways to control these pests. Chemicals still had their place, and probably always will. However, something had to take the place of those products that were no longer effective or were too dangerous for continual use. Among the potential solutions examined was the use of the "white hats" versus the "black hats". Biological control was reborn.

Over the years numerous insects became the targets of this method of control. Sometimes the white hats won; sometimes the black hats prevailed. But it was always a fascinating battle!

One of the most recent examples of the spectacular use of insects to control insects was in the cereal leaf beetle war. The beetle was an invader from Europe. It arrived on the shores of the Great Lakes and immediately attacked our grain crops. Its huge appetite for wheat, oats, and barley made it a severe threat to crops in the Northcentral U.S.A. The first line of defense was to eradicate the beast with chemicals. This failed. Then containment of the enemy by barriers was tried. This strategy was ineffective. The beetle kept on increasing and causing serious damage.

Finally, when everything else failed, the use of beneficial insects to do battle was considered. Because the invader beetle was from Europe, eyes turned to the Old World for a solution. Surveys were made all over Europe by U.S. Department of Agriculture scientists of the Agricultural Research Service. Their task was first to find the beetle and then to collect any insect attacking it.

After several seasons of research with the pest and its complex of natural enemies in Europe, four species of tiny wasps were shipped to the United States to do battle. They were studied, mass reared, and distributed throughout the invaded grain fields by the Animal and Plant Health Inspection Service of the USDA.

The wasps worked. Within a very short time the beetle was defeated, although not annihilated, and the grain fields were saved. All this cost about a million dollars. The annual savings from this program is estimated at many millions of dollars, not only in the cost of chemicals not applied, but in the reduction of pollution, and other environmental considerations.

Another insect pest brought to its "knees" by the concerted action of the good guys was the alfalfa weevil. The weevil had arrived in the western United States very early in this century. It raised havoc with the alfalfa crops of the region. A parasitic wasp from Europe was found and introduced but the degree of control was not spectacular.

In the early 1960s, a new wave of the weevil innundated the alfalfa fields of the eastern part of the country. The effects were disastrous. Farmers lost thousands of acres to this voracious beast. Some surrendered and stopped growing alfalfa. They turned to corn and other crops for feed. Others refused to bow to the beast but had to subject their crops to tons of insecticides just to stay in business. This approach just was not realistic.

Again the agricultural community sought out biological control for an answer. Again the answer was there. The scientists at the U.S. Department of Agriculture's European Parasite Laboratory in the suburbs of Paris went into action. Europe was combed and several species of tiny wasps were found living in Europe as parasites of the alfalfa weevil. These creatures in the "white hats" were collected, studied, and shipped to the United States.

The wasps were spectacular in their action. Within a few years after their release the use of chemical pesticides against the weevil was reduced by ninety percent in some areas. Farmers returned to growing alfalfa for their livestock. The weevil was reduced to a non-pest status. Periodically, the beetle still rears its ugly head when, for one reason or another, the life cycle of the wasps gets out of synchronization with that of the weevils.

Although peace had returned to the green fields of alfalfa, another enemy waited in the wings. A small European fly found its way to the luxuriant plantations of this essential crop. In no time at all, this leafmining fly had replaced the weevil in the Northeastern States as the most serious pest of alfalfa. This new insect produced an army of maggots from eggs laid between the surfaces of the plant leaves. This ravenous horde devoured the substance of the leaves on the plants. The mined and dead leaves dropped off the plant. The fields became brown. At harvest time the farmer was faced with a crop of bare stalks instead of nutritious green plants.

The farmer turned to chemicals to combat this scourge. But, the poisons released to kill these new black-hatted villains also wiped out the white hatted insects standing guard over the defeated alfalfa weevil. Things looked bleak. Not only did the farmer face a new threat from the fly, the old nemesis was regaining strength.

Again, Europe became the source of the solution. Exploring entomologists found species of wasps living on the weevil in all parts of the continent. Of the 14 different species sent to the United States, only three became established in the New World. Two of these three species formed their own army and soon dispatched the leafminer. Chemical applications were drastically reduced. The rejuvenated weevil parasites regained the upper hand and the alfalfa weevil was again under control. Calm returned to the fields.

Similar battles and full scale wars are in progress in many parts of the United States and throughout the world. Yet, people are unaware of most of them.

Scientists who have devoted careers to classical biological control more or less shoot themselves in the foot every time they successfully control a pest. If the success is complete the natural enemies become a part of the environment. They take up permanent residence and maintain continual surveillance of the pest. If the invaders shows signs of becoming problems the natural enemies step in and take command. The scientist is no longer involved. By then the interest on the part of the officials and the public has turned to other things. Almost before they know it the scientists are being asked "What have you done for us lately?"

What have we done? We have done quite well, actually. In recent years the biological control of just five insect pests — the cereal leaf beetle, the alfalfa weevil, the alfalfa blotch leafminer, the Rhodesgrass mealybug, and the pea aphid — has resulted in an annual savings conservatively estimated to be about $130.5 million. Added to this benefit is a considerable reduction in environmental pollution. However the benefits are calculated, the progam was well worth the effort and expense.

How much did it cost to declare war on these pests and initiate the battles? A rough estimate was about $1 million per project. Some projects cost less, others more. But the investment did pay off!

What does the future hold for the field of Classical Biological Control? That depends upon how much the state and federal governments wish to invest.

This science must be made more attractive, the future more promising. There is a great deal of talk about supporting biological control but it is the actual funding that will make the difference.

Classical biological control with insects normally does not result in a marketable product. Consequently, tremendous investments from industry are not to be expected. Most of the introduced natural enemies cannot be stockpiled. As soon as they are declared safe for release after importation they are put into the field to do their task. If they are successful the pest is defeated. Then, most of the agents of the victorious army die off without attacking other insects, animals, or man. But a small number of the winning species remain, with little if any help from man, ready to renew the attack if the pest begins to rebuild its forces. If a project is successful it is self-sustaining and there is nothing to sell. At present the name biotechnology looms large on the horizon and, in some cases, threatens to engulf classical biological control. Hopefully, the pursuit of this fascinating, but still nebulous, area of research does not dry up the financial wells that nourish the classical approach to biological control.

Man will always need the white hats as long as there are villains. The United States has been invaded by many pests over the years. Furthermore, it is estimated that at least one serious villain of the insect world invades our country every three years. As a result there is a tremendous backlog of pest insects just waiting to be controlled. For some, the agents in white hats that are called in effectively do their job. The villain is reduced to a non-pest status and, for the most part, the problem is solved. For others, the natural enemies are ineffective and other control methods must be considered.

With the speed of international travel and transport we can be certain that the bad guys will continue to find their way to our shores. When this occurs, the scientist eventually must return to the fields, forests, deserts, and mountains in all parts of the world to discover and collect the "good guys of agriculture" to help protect our crops and environment from the uninvited immigrants.

Ladybird beetles

Robert D. Gordon

Ladybird beetles are members of the family Coccinellidae, or "coccinellids" and are the most popular and among the best known of all insects. They have been objects of general popular interest for centuries because of the bright, contrasting red and black colors of many species. They also tend to either feed or hibernate in large numbers and are commonly found in or around dwellings seeking winter shelter, hence are readily visible.

Popular interest in the ladybird (which in Europe is *Coccinella septempunctata*, the seven-spotted Ladybird) goes back at least to the fifteenth century and probably much farther. The ladybird is usually dedicated to the Virgin Mary; in Scandinavia it is called Nyckelpiga, our Lady's Key-maid, or Jung-fru Marias Gulhona, the Virgin Mary's Golden-hen. In Germany it is Frauen or Marien-Kafer, Ladybeetles of the Virgin Mary, and in France it is known as Betes de la Vierge, Animals of the Virgin. In England it is Lady-bird, Lady-bug, Lady-fly, May-cat, etc.

In Europe, particularly in Germany where most of the superstitions connected with it are supposed to have originated, the ladybird is always connected with fine weather.

From Vienna a superstition connected with the ladybird's ability in this regard:

Little birdie, birdie
Fly to Marybrunn
and bring us a fine sun

Marybrun being a place near the city with a miracle-working image of the Virgin that is thought to send good weather to Vienna. From the Elbe marshes comes a similar request:

May-cat
Fly away
Hasten away
Bring me good weather with you tomorrow

Northern Germany gives us a request based on the belief that the ladybird can foretell the harvest year; if the spots exceed seven, grain will be scarce, if there are fewer than seven, there will be an abundant harvest:

Maerspart, fly to heaven
Bring me a sack full of biscuits, one for me, one for thee
for all the angels one

In northern Europe it is thought to be lucky when a young girl sees the ladybird in spring, she lets it creep around her hand and says, "She measures me for wedding gloves". When it flies away the direction it takes is important because it signifies from what direction her sweetheart will come. England provides us with this rhyme:

This Ladyfly I take from off the grass
whose spotted back might scarlet red surpass
Fly, Ladybird, north, south, or east or west,
Fly where the man is found that I love best
He leaves my hand, see to the west he's flown,
To call my true-love from the faithless town

Also from England (Norfolk) comes a similar wish in verse:

Bishop, Bishop Barnabee
Tell me when my wedding be:
If it be tomorrow day,
Take your wings and fly away!
Fly to the east, fly to the west,
Fly to him that I love best.

Some superstitions have existed about the ladybird that don't appear in verse, such as the Ladybird as a cure for measles and colic, or as a cure for the toothache when specimens are mashed and put into a hollow tooth. This latter use of the ladybird was tried by an English gentleman in 1859 and reported thus: "I tried this application in two instances, and the tooth-ache was immediately relieved; but whether the remedy, or the faith of the patient, acted therapeutically, or the tooth ceased aching of itself, I confess I do not pretend to know".

Ladybird beetles are generally thought of as beneficial insects, predators of plant pests; this is true for the most part, particularly in temperate regions. In tropical regions, however, many are plant feeders, some of economic importance. A few plant feeding ladybirds occur in temperate regions, the Mexican bean beetle in the eastern United States being the prime North American example.

Historically, the beneficial ladybirds have been considered predators of aphids or "plant-lice" and, in fact, many are, but many more species make a living devouring scale insects and mealy bugs. One group of ladybirds has become adapted to feeding on mites, and only on mites; still another group feeds only on fungal spores and hyphae. Certain species have even developed the ability to survive on plant pollen when aphids are scarce. The most bizarre feeding strategy known for the entire coccinellid family is possessed by a group of South American species. Members of this group are actually cannibals because they feed exclusively on the larvae of another group of ladybeetles!

Knowledge of the feeding habits of beneficial ladybirds eventually led to the first attempts to use them for biological control. Most serious plant pests are not native to the area where they have become pests. They have been introduced from other parts of the world, nearly always by man, without bringing along the natural parasites and predators that kept them under control in their native habitats. In the words of an American entomologist in 1855; "We have received the evil without the remedy". Therefore it is logical to attempt to control introduced pests by determining from whence they came, then locating and introducing their natural enemies.

The cottonycushion scale, a serious pest of citrus in California, precipitated the first attempt to introduce foreign parasites and predators into North America. In 1888, a man named Albert Koebele was sent to Australia to obtain natural enemies. He sent back to California several species of ladybirds, among which was the now famous "vedalia" beetle. This species proved to an immediate and spectacular success, in fact it has been credited with saving the California citrus industry. This success precipitated a wave of coccinellid introductions that included 46 Australian species between 1891 and 1892. Very few of these became established, and the interest in predaceous ladybird beetles waned in favor of parasitic wasps and, later, pesticides. It was not until the 1960's and 1970's that coccinellids were again introduced in significant numbers with several useful establishments resulting. Available records show that approximately 179 species have been intentionally imported into North America and a few species imported by accident, probably as hitchhikers on ships or planes. Ten of these accidental immigrants have become established along with 16 of the intentionally introduced species, a total of 28 foreign ladybird beetles are currently established in North America.

The future of biological control, for many years in jeopardy because of almost total reliance on pesticides, once again looks promising. The past two decades have seen a strong realization of the harmful effects of pesticides followed by restrictive pesticide legislation; the result being revived interest in biocontrol. Experience gained over the last 90 years, coupled with advances is laboratory methods, rearing techniques, and increased funding, will enable scientists to utilize predators and parasites for pest control far more efficiently than ever before. The full potential of ladybird beetles as biological control agents may finally be fully realized. Recent discovery in the United States of the Russian wheat aphid, a serious pest of small grains, will test this prediction. Several species of coccinellids are being studied, along with some parasitic wasps, for control of the aphid which is the first serious new immigrant pest to enter North America in many years. Success in achieving Russian wheat aphid control will virtually ensure government funding and other support for biological control for years to come.

The Monarch Butterfly — Our National Insect

Douglas W. S. Sutherland

Why have a national insect? Our national bird is the bald eagle, and our national flower is the rose. Insects constitute about 80 percent of the world's animal species, and they are critical to the ecological balance of our planet. Their great numbers and diversity play a vital role in our daily lives. It is therefore fitting that we have a national insect. And among the insects, the grace, beauty, and uniqueness of the native monarch butterfly (*Danaus plexippus* (Linnaeus)) make it an excellent representative of our natural wildlife heritage.

The Entomological Society of America (ESA) hoped that the citizens of our nation would adopt a national insect after the membership of the ESA first voted for the monarch butterfly as its choice.

Monarch butterfly history

The Indians of North America were undoubtedly the first to observe and illustrate the monarch butterfly (Sullivan 1987). And in the 1880's Chief Sitting Bull was photographed with a monarch mounted on his hat (Capp 1975).

The oldest known specimens of this unique species were found in the collection of James Petiver, an English pharmacist. One of these was collected in Maryland between 1698 and 1709 and preserved between sealed sheets of mica (Wilkinson 1969). *Danaus plexippus* was described and named from Pennsylvania specimens in the 1758 10th edition of *Systema Naturae* by the Swedish botanist, Carolus Linnaeus. This text is accepted by zoologists as the beginning of the modern classification of animals.

The monarch butterfly is frequently mentioned in the early naturalist literature of the United States. Catesby (1734-43) in the southeastern U.S. and Kotzebue (1821) in California are among the authors who published books with good illustrations of the monarch. Over the years it has also been called the archippus butterfly, King Billy, milkweed butterfly, and wanderer (Harris 1862; Sutherland 1989).

Lepidopterist Fred Urquhart (1987) states that the early settlers of Colonial America were impressed by this striking insect and gave it the name "monarch" in honor of "King William, Prince of Orange, stateholder of Holland, and later King of England." Therefore the name itself is part of the cultural heritage of our nation.

Overwintering sites, the spectacular "butterfly trees" of central and southern California, have been known for more than 100 years. Urquhart has studied monarch butterfly habits since 1937 (Urquhart 1960. 1987), yet it was only in 1975 that he and naturalist Ken Brugger discovered its Mexican overwintering locations (Urquhart 1976). This butterfly has been studied and admired by generations of school children. For instance, each fall the youngsters of Pacific Grove, California welcome the return of the monarch butterflies with a parade. The monarch was designated the state insect of Illinois in 1975, the state butterfly of Vermont in 1987, and the state insect of Alabama in 1989.

Biology

How does the monarch butterfly live and survive? A generation begins when the females lay their eggs on milkweed. The caterpillars hatch from the eggs, feed, grow, and molt five times. Fully grown two-inch-long, yellow, black, and white-banded caterpillars, each with two pairs of black filaments, attach themselves to their food plant or nearby object with silk threads. Then they transform into the resting stage within striking, glossy, lime-green and gold-spotted chrysalis, each marked with a black and gold band. The metamorphosis to the butterfly stage takes place within this chrysalis.

The spectacular black and burnt-orange adults emerge about six weeks after the eggs are laid. The butterflies, with about a four-inch wingspan, flit about, glide, cruise, partake of social, prenuptial, and speed flights, mate, and lay eggs (Urquhart, 1987). They feed on the nectar of milkweed and other flowers and help in pollinating them. They are most common in open fields and marshes where milkweed, the main larval food-plant, grows.

Each year there are four to six generations in the south and one to three in the north. The monarch butterfly is found in southern Canada, throughout the United States, Mexico, and Central and South America. In about 1850 the monarch became established in Hawaii and other Pacific islands. But it is in the United States that it is best known and appreciated.

With the end of each season, the monarch butterfly is dependent for survival on its unerring and still incompletely understood ability to migrate south. Overwintering sites are in coastal California, south central Mexico, and possibly still undiscovered locations in Florida or the Caribbean islands. This migration phenomenon is unique in the insect world (Zahl 1963; Urquhart 1987).

Migration, overwintering, and survival

In August the adults of the last generation of each year prepare to migrate by storing fat for the journey of 1,800 miles or more. Many hundreds of volunteers have been involved for decades in tagging monarch butterfly adults throughout its range and compiling the data to verify the incredible journey of this insect (Urquhart 1987; Urquhart & Urquhart 1990).

Entomologists believe that the shortening days and cool nights trigger this migratory urge. The butterflies can be seen following the rivers and shorelines and flying over mountains and bodies of water. Their strong, graceful flight and intermittent "sailing" at altitudes of 3000 feet or more take advantage of the tail winds as they work their way south.

Overwintering sites provide high humidity and protection from the elements at temperatures that allow the butterflies to remain inactive and thus to conserve food reserves. Festooned on trees in protected mountain forests and coastal groves, the butterflies wait out the winter by the countless millions. During short warm periods they become active, seek puddles of water, and drink to keep from becoming dehydrated.

Approximately one percent of the overwintering generation survives to begin the northward journey in March as night temperatures moderate and days become longer. Few if any are believed to make it back to their individual birthplaces. Instead, the females lay eggs and die along the way. By late April the monarch butterflies have leap-frogged to southern Illinois and New York, and by May they have arrived in the northern United States and southern Canada.

Ecology

There are many more unusual facts about the monarch butterfly. For instance, it is distasteful to birds that attempt to eat it. Scientists Lincoln and Jane Brower and others have studied this phenomenon and found that the stored glycoside chemicals that protect the monarch butterfly come from the milkweed plant (Brower, *et al.* 1988). This protection is so important to survival that, over time, a very edible butterfly, the Viceroy, has come to mimic the monarch and is, therefore, also protected from birds and other natural enemies.

Conservation of overwintering sites

There is a combined international effort to maintain and preserve the remarkable monarch butterfly for future generations to enjoy.

The Monarch Project of the Xerces Society in the United States seeks to educate citizens about the monarch butterfly's uniqueness (Pyle 1989). Better understanding of the navigational mechanisms and in plotting migration routes of the monarch is being sought. The Xerces Society is also obtaining easements for the less than 45 known overwintering sites in California and is coordinating efforts with Monarca A.C., a group of concerned naturalists in Mexico, to preserve overwintering sites there. The World Wildlife Fund has provided financial support to Monarca A.C. for its conservation efforts. The Mexican government has set aside a number of these overwintering locations as "ecological preserves," which are of critical importance for survival. Furthermore, in 1983 the International Union for the Conservation of Nature and Natural Resources designated the monarch butterfly overwintering sites a "threatened phenomenon." Other organizations, such as Friends of the Monarchs, in California, are also active in monarch conservation efforts.

Seldom has there been such coordinated interest, concern, and support for the conservation and survival of one of our natural wonders.

The status of "Insect USA"

The idea of a national insect was conceived by entomologists Jim Johnson, Paul Opler, and Doug Sutherland on a plane trip between Reno, Nevada and Denver, Colorado following the 1986 National Meeting of the Entomological Society of America. The then upcoming 1989 Centennial of ESA was on our minds and this seemed a logical project to undertake to increase the environmental and entomological sensitivity of our citizens. Our choice was the monarch butterfly. Following submission of the proposal to the ESA Governing Board and acceptance of the idea by the membership in 1987, the project evolved to become nationwide interest. In 1988 the project expanded beyond ESA to include the original 11 supporting organizations listed in a promotional brochure (Sutherland, et al. 1988).

However, it was not until September 27, 1989, when Congressman Leon E. Panetta from Santa Cruz, California (in the heart of the area where western monarchs overwinter) introduced House Joint Resolution 411 designating the monarch butterfly as the national insect, that the campaign began in earnest.

In addition to Congress and other organizations, countless individual garden club members, environmentalists, teachers, and others worked to bring about the passage of H.J. Res. 411. For students this is a lesson in insect ecology and an opportunity to be a part of the democratic process and study our government in action.

The monarch butterfly has received wide coverage in newspapers and elsewhere. A National Wildlife Federation article in Ranger Rick magazine elicited close to 1000 responses (Anon. 1989). The supporting resolutions were passed by a number of the garden federations and clubs.

Monarch motifs are available on T-shirts, in porcelain, as note cards, on cups and wine bottles, as ties, and other articles.

There seems no end to the scientific articles on the monarch butterfly and the popular books on insects and nature study, almost all of which contain some reference to the monarch butterfly. Trips to the overwintering sites in California and Mexico are available from travel agencies.

Monarch butterfly displays were shown at garden club and other meetings for Earth Day '90 celebrations. It is also a subject for study or display at insect zoos, nature centers, and museums around the country.

Because the United States is rapidly becoming a nation of urban dwellers, it is good to remind citizens of the value of conserving our natural heritage, the importance of a healthy environment for the survival of all living things, and the vital role that insects play in the ecology of our daily lives. A national insect provides a mechanism for this outreach to the citizens of the United States.

Acknowledgements to:
Gordon Gordh, James Johnson, Paul Opler, Kevin Steffey, and Frederick Santana of the ESA National Insect Subcommittee; J. G. Rodriquez, Chairman of the ESA Centennial Committee; and Julian Donahue, John Kliewer, Lloyd Knutson, R. K. Walton, and others who contributed significantly to this document.

References

Anon. 1989. Insect U.S.A. Ranger Rick. National Wildlife Federation, Washington, DC. 23(10):41.

Brower, L. P. *et al.* 1988. Exaptation as an Alternative to Coevolution in the Cardenolide-Based Chemical Defense of Monarch Butterflies (*Danaus plexippus* L.) against Avian Predators. Chapter 15. pp 447-475. In Chemical Mediation of Coevolution. Amer. Inst. Biol. Sci., Washington, DC.

Capps, B. 1975. The Great Chiefs. Time/Life Books, Alexandria, VA. pp 194-195.

Catesby, M. 1734-43. A Natural History of Carolina, Florida and the Bahama Islands. vol II. plate 88.

Harris, T.W. 1862. A Treatise on Some of the Insects Injurious to Vegetation (3rd ed). Willliam White Printer, Boston. pp 280-281.

Kotzebue, O. von. 1821. A Voyage of Discovery into the South Sea and Bering's Straits for the Purpose of Exploring a North-East Passage, Undertaken in the Years 1815-1818 [a translation]. London. III:451. Fig. 14.

Pyle, R. M. 1989. Monarch Butterflies: messengers for invertebrates. Wings. Xerces Society, Portland, OR. 14(2):11.

Sullivan, S. 1987. Guarding the Monarch's Kingdom. International Wildlife. 17(6):4-11.

Sutherland, D. W. S. and D. Brassard. 1989. EPA List of Insects and Other Organisms. Office of Pesticide Programs, USEPA, Washington, DC. 167 pp.

Sutherland, D. W. S. *et al*. 1988. National Insect: the monarch butterfly. Entomol. Soc. Amer. Lanham, MD. 8 pp.

Urquhart, F. G. 1960. The Monarch Butterfly. Univ. of Toronto Press. Toronto, Canada. 361 pp.

Urquhart, F. G. 1976. Found at Last: the monarchs winter home. Natl. Geog. 150(2):161-173.

Urquhart, F. G. 1987. The Monarch Butterfly: international traveler. Nelson-Hall. Chicago, IL. 232 pp.

Urquhart, F. G. and N. Urquhart. 1990. Insect Migration Studies. Annual Rept. Univ. of Toronto. Toronto, Canada. 27:22 pp.

Wilkinson, R. S. 1969. The Oldest Extant Specimens of North American Lepidoptera. The Michigan Entomol. 2:46-47.

Zahl, P. A. 1963. Mystery of the Monarch Butterfly. Natl. Geog. 123(4):588-598.

The Honey Bee Deserves to be Our National Insect

H. Shimanuki

In the midst of Congressional discussions on the drug problems, star wars, education, flag burning, and taxes, comes as a hotly debated topic: the National Insect. The most beneficial insect known to man, the honey bee deserves to be our national insect. Let me present some of the facts about this insect.

The Immigrant

The honey bee is not native to the New World. The honey bee can trace its beginnings in this country to the early 1600s when it was first imported into Virginia. How many of us have a genealogy that can be traced that far back? The residency requirement for the President of this country is that the individual be a U.S.A. citizen by birth. If there is no requirement that the U.S.A. President's forebears be native Americans, then perhaps there should be no such requirement for the selection of our National Insect?

The honey bees were imported from many countries to produce a sweetener, honey, for the settlers who had great difficulty obtaining sugar. The strange insects were foreign to the native Indians who had no term for honey bees, so they called them "white man's flies". U.S.A. honey bees today are a mixture of many races, originating from many countries. They are truly representative of America's "melting pot" in the insect world; countries include Germany, Italy, Yugoslavia, Russia, Great Britain, Morocco, Tunisia, Syria, Cyprus, and many more.

The honey bee society reflects our democratic values as well; there are no dictators. Although there is a member of this society that is called the queen, she really has little to say about the fate of the colony or even herself. The queen is simply the egg-laying machine. She lays upwards of 1,500 eggs a day during the peak of the season. She does not stop to find food or groom herself. Her attendants make certain that she is fed and cleaned. By controlling the queen's diet, her space in which to lay eggs, and which eggs will be allowed to develop into adults, the worker bees are the ruling class and therefore can be likened to the American working class, the voters.

The Work Ethic

At its peak, the average colony may have about 40,000 adult workers. This caste is well-named as these bees truly work. Development from egg to emergence takes 21 days. Upon emergence they immediately begin to work. Their first task is to clean the cell in which they were reared. After this task, the typical worker consumes large amounts of honey so it can produce beeswax and construct combs for food storage and brood production. In addition, these same workers consume pollen to produce royal jelly and worker jelly. This then becomes their second task, care and feeding of the queen and brood.

Working for the Benefit of the Colony

When workers attain the ripe old age of 21 days they leave the hive to forage for food. The worker bees collect pollen, nectar, propolis, and water. Worker bees literally work themselves to death. Foraging for food results in encounters with man and some of his weapons of destruction — automobile windshields and pesticide sprays. In addition, the foragers may encounter other problems: insects, birds, bats, and toads.

The Patriot

If the worker overcomes all these devastators, she will then die a normal death, unable to fly because of frayed and tattered wings, the result of intensive work. Once worker bees attain the age of 18 days they may be called upon to perform guard duty. These guard bees will give up their lives by stinging any intruder and will subsequently die from the loss of their sting. The intruders into the hive could be other insects, wild animals, or man.

Preserving Limited Resources

In addition to queens and workers, there are the drones, the males of the honey bee colony. The primary function of the drones is to mate with the queens. If they are successful in mating, they die before they can return to their nests. The males are truly defenseless, they have no stings. Furthermore, drones cannot even feed themselves. As winter approaches and there is no incoming food, this is a signal for the workers that drones are no longer needed for mating until next Spring. Consequently, drones are pushed out of their colony by their sister bees.

Starting a New Home

In the Spring, when food is available and the colony population is growing faster than the space for food storage and brood production allows, the colony begins to prepare for swarming. The queen makes no decision here either, the workers by an "act of succession" or a "democratic vote" decide to raise new queens in preparation for swarming. When the new queens emerge or just prior to their emergence, the old queen with a portion of the colony leave to find a new nest.

Before the swarm leaves the hive, it sends out scout bees to look for possible nesting sites. As if by consensus, the swarm surveys locations identified by the scout bees until it finds a suitable home. The nest is usually a protected cavity, close to forage and water, easily defended from predators and with adequate sun exposure.

Protecting Our Environment

Some of the benefits of honey bee pollination are frequently overlooked because they occur without the intervention of man. For instance, honey bees pollinate such things as sea oats and other plants that help maintain sand dunes on our beaches. Honey bees also contribute to our wildlife by pollinating some of the fruits and berries that are an important part of the food chain of birds, bears, and other animals.

Air Pollution Monitor

Honey bees are being used to monitor air pollution and hazardous wastes. Unlike high technology devices that require big budgets to operate, the honey bee colony can survive in remote areas; it needs no electricity; it is virtually maintenance-free; and it is self-propagating. As they forage, honey bees collect samples of nectar and pollen as well as sampling the air and water from an area covering 12 square miles. Yes, honey bees are doing their share in helping to monitor air pollution. By analyzing the effect of pollutants on the egg hatch, pollen, honey, and the adult bees, it is possible to detect industrial pollution in remote areas of the country.

Products of the Hive

Honey: There are probably thousands of ways honey has been used through the centuries. Honey is not merely a sweetener. Its uses can be traced to biblical times and in literature, references are made to honey as "the nectar of the gods." Honey is used in baking because it helps pastries retain moisture, and equally important, it also contributes to the flavor. For instance, how would you like to eat baklava made with sugar syrup, or have a nice hot biscuit with sugar syrup? Because of the limitless variety of nectar sources that the honey bee visits, there are infinite possibilities of honey flavors. Honey is used in pastries, candies, salad dressings, jams and jellies, ice creams, barbecue sauces, and in hundreds of other foods. In some countries honey is used to make a wine, mead, and in Africa, honey is used to make beer.

Honey has been used in hair shampoos, skin lotions, facial masks, curing tobacco, and preparing briar-root pipe bowls. In some cultures, honey is used in formulating home remedies for minor ailments, and as a dressing for minor wounds and burns.

Beeswax: The honey bee is a wax factory. Worker bees manufacture beeswax using honey as a raw product. About seven pounds of honey are required to produce one pound of beeswax. The beeswax is used for building combs for their nest - a structure for storing food and serving as a nursery.

A list of 1,001 uses for beeswax could be made. Most beeswax is used in the cosmetic, candle making and the beekeeping industries. Beeswax is used in facial creams, lipsticks, and other beauty aids. Some churches still require the use of beeswax in making candles for their services. Beekeepers also trade their unbleached beeswax cakes for beeswax foundation that honeybees use as a starter for comb in the hive. There are other uses for beeswax, such as furniture polish, lubricants for artillery shells, coating leaders for fly fishing, pre-treating nails before driving them into some woods, ironing etc.

Pollen: Bee-collected pollen is found in many health food stores. It is used as a diet supplement by some individuals. Because pollen is obtained from many plant sources, its protein content could vary from 8 to 40 per cent on a dry weight basis. In addition, pollen contains some minerals and vitamins.

Royal jelly: Royal jelly is the primary food fed to queens throughout their lives. In some cultures, it is believed by some to have a medicinal value. The cosmetic industry also uses royal jelly in some of their formulations.

Propolis: Sometimes this product is called "bee glue" because it is used by the honey bees to seal small openings in their nest. Propolis is of plant origin and foraging honey bees must collect it from trees. In parts of Europe, propolis is used to make a tincture to treat minor cuts. Others use propolis as a lozenge for sore throats. Propolis has been shown to possess some antimicrobial properties but because it is collected from various trees, its efficacy may vary according to sources.

Gourmet food: In some countries, honey bee brood is eaten; bee brood is crushed while extracting honey and thus consumed with the honey. If you are squeamish about eating them raw, try cooking them in deep fat. I have been told that they have a nutty flavor when prepared in this manner. If this does not suit you, you can always try chocolate covered bee brood.

Bee venom: Bee venom is collected by a few beekeepers to sell to the pharmaceutical industry. It is used to desensitize those allergic to bee stings and in some countries, it is used to treat arthritis.

Wax moth: If pesticides, diseases, or starvation destroys the colony, the greater wax moth could invade the nest and destroy the combs. If this happens, do not despair, the wax moth larvae is an excellent fish bait. Take some time off from keeping bees and go fishing.

Conclusion

The honey bee is indispensable to agriculture. Because of their importance, some have called the honey bees "the angels of agriculture." The honey bees are industrious, just like the immigrant American that made this country great.

The honey bee has been named as the state insect by a number of states in recognition of its contribution to American agriculture. In addition, the United States Postal Service has honored the honey bee by issuing two stamps in its honor.

The next time you dine and give thanksgiving, remember the honey bees. One-third of what Americans eat is directly or indirectly the result of honey bee pollination. Without the honey bees, food costs would soar and we would not have the variety that we enjoy in our diets.

Don't you think that for all the preceding reasons the honey bee deserves to be our National Insect?

Acknowledgments

I would like to thank my wife, Vivian, for proof-reading this manuscript and making many helpful suggestions, and to the honey bees who made it possible for Vivian and me to meet.

Suggested Further Reading

The ABC and XYZ of Bee Culture. 1989. Edited by R.A. Morse. A.I. Root Company. Medina, Ohio.

Bromenshenk, J.J., S.R. Carlson, J.C. Simpson, and J. Thomas. 1985. Pollution monitoring in Puget Sound with honey bees. Science 227: 632-634.

Dietz, A., R. Krell, and M.S. Brower. 1982. Pollination and our seashores. Proceedings 10th pollination Conference, Southern Illinois University, July 1982, Carbondale, Illinois. pp. 57-66.

The Hive and the Honey Bee. 1975. Edited by Dadant and Sons. Hamilton, Illinois. 740 pp.

Honey, A Comprehensive Survey. 1975. Edited by E. Crane. Heinemann, London. 609 pp.

The Illustrated Encyclopedia of Beekeeping. 1985. Edited by R.A. Morse and T. Hooper.

McGregor, S.E. 1976. Insect Pollination of Cultivated Crop Plants. USDA Handbook No. 496. 411 pp.

Robinson, F.A. 1982. Pollination and wildlife habitats. Proceedings 10th pollination Conference, Southern Illinois University, July 1982, Carbondale, Illinois. pp. 67-76.

Robinson, W.S., R. Nowogrodzki, and R.A. Morse. 1989. The value of honey bees as pollinators of U.S. crops. American Bee Journal 129:411-423, 477-487.

Take Two Cicadas and Call Me in the Morning

Gene R. Kritsky

If you hear a ringing sound in your ears, clinically called tinnitus, then maybe you should boil some cicada skins and drink the extract. Although this sounds strange, it is just one of the uses of cicadas in traditional Chinese medicine.

Chinese medicine developed during several centuries of cultural isolation. The oldest Chinese records of diseases and possible treatments have been found carved on bones and tortoise shells dating from 1600 to 1100 B.C. Even in these early writings are found prescriptions for the cure of many diseases. During the Ming Dynasty (1368-1644) traditional Chinese medicine reached a peak in its development with the 168-volume, *Prescriptions for Universal Relief* which included 61,739 prescriptions for the treatment of various internal diseases. Herbal medicine was quite important in the Chinese treatment of disease, but insects also became essential ingredients for several prescriptions (Jizong & Zhu 1985).

Traditional Chinese medicine, although founded in the past, still enjoys popularity in China. Visitors to some of China's most modern hotels can purchase in hotel drugstores dried cicada nymphs, antlers, and ginger roots. I recently had the opportunity to visit the People's Republic of China as part of a scientific exchange among entomologists, and I visited traditional Chinese drugstores in Xian, Chongqing, and later in Hong Kong. With the aid of an interpreter, I was able to purchase several insects that were commonly sold as part of traditional prescriptions.

A visit to the local Chinese drugstore brings an unusual array of odors and sights. The patient takes the prescription to a counter where he or she pays for the various ingredients. The receipt is then taken to the drug counter, which stands in front of a wall of drawers filled with various herbs, minerals, and insects. The typical prescription includes several different ingredients, which are weighed out in small hand-held pan balances. The various items are then packaged together into a large mass, which is later boiled and strained. The patient drinks the extract to obtain the desired effects of the various ingredients.

The Chinese use of insects and arthropods is quite varied. I was able to purchase or inspect dried cockroaches, blister beetles, maggots, silkworm larvae, cicada exuviae, cicada nymphs, and adults. I have also found recipes for Chinese potions using silkworm excrement, amber, honey, mole crickets, mantid oothecae, dried crickets, scorpions, and centipedes. My inquiries regarding what ailments could be treated by various insects yielded a wide range of replies. Meloid beetles were said to be used for dermatological problems. Even the Natural History Museum in Beijing had a display of meloid beetles that included a bottle of pills apparently containing meloid extract. The pharmacist in the largest traditional Chinese drugstore in Chongqing recommended cockroaches for stroke victims, and a medical text included silkworm excrement for the treatment of typhus and stomach troubles.

The medicinal effectiveness of the insect ingredients has not been studied thoroughly. Many of the herbal ingredients in Chinese prescriptions are known to have significant effects on the body. For example, licorice root will lessen pain, aid in the treatment of ulcers, and increase the effects of other ingredients (Richardson & Stubbs 1978). Clausen (1954) discussed some of the clinical effects of insects used as a medicine in other cultures. For example, meloid beetles taken internally can affect the kidneys. However, not all of the insect ingredients produce a physiological response. Some of the prescriptions that incorporate insects may function by inducing psychosomatic cures. One example might be the treatment for bedwetting in children. The prescription is to simmer together seven ginkgo seeds beaten with their shells, 9 g of mantid ootheca, and 9 g of *Lindera* (spicebush) root. The child should drink the extract twice a day for five days and then every other day until he is cured. Because ginkgo seeds have a characteristic unpleasant odor when they are still in the fleshy integument, it is more likely that the child, repulsed by the odor, is given a strong incentive to stop bedwetting (Jizong & Zhu 1985).

The traditional Chinese treatment of hypertension presents a typical case study of how Chinese philosophy explains the cause of an ailment and how insects are combined with herbs to treat a disorder. According to traditional Chinese medicine, hypertension is caused by an imbalance of yin and yang. Yin and yang are philosophical concepts that have important relationships to everything from the origin of the universe to the treatment of diseases. They apparently were first applied to medicine during the Spring and Autumn period (770-476 B.C.). Yin and yang each refer to different parts of the human anatomy. For example, yin denotes the viscera, blood, and abdomen, whereas yang constitutes the bowels, vital energy, and the back. Hypertension is thought to be caused by too much yang (vital energy) and too little yin (blood), and treatment requires bringing the two back into balance (Jizong & Zhu 1985).

Shi Jinmo (1881-1969), a noted Chinese physician, had a complex prescription for the treatment of hypertension and related symptoms. A 61-year-old woman suffering from hypertension, dizziness, tinnitus, vomiting, and frequent urination was treated by Jinmo with two prescriptions given in two stages. The first prescription included 30 g each of magnetite, sea-ear shell, and *Eucommia* bark; 15 g each of hematite and *Achyranthes* root; 10 ginkgo seeds; 9 g each of *Polygala* root, self-heal spike, chrysanthemum flower, *Ophiopogon* root, *Tuckahoe*, *Tribulus* fruit, and flattened milk-vetch seeds; 6 g each of *Inula* flower, swallow-wort root, and grass-leaved sweet flag rhizome; 5 g each of cicada exuviae and *Gastrodia* tuber. The patient took four doses of this prescription and claimed to feel better. The second prescription consisted of a mixture of 60 g each of *Tribulus* fruit and *Eucommia* bark; 30 g each of *Ophiopogon* root, *Achyranthes* root, sea-ear shell, *Cibot* rhizome, *Rehmannia* root, *Arborvitae* seeds, wild *Jujuba* seeds, mulberry leaves, chrysanthemum flower, *Eclipta*, privet fruit, *Polygala* root, grass-leaved sweet flag rhizome, flattened milk-vetch seeds, and *Tuckahoe*; 15 g each of *Euodia* fruit, chuanxiong rhizome, Chinese gentian, cicada exuviae, *Amommum* fruit, and licorice root. This mixture was powdered, mixed with honey, and made into pills each having a mass of 9 g. The patient took a pill, sometimes called a bolus, each morning and evening with warm boiled water and completely recovered. The various ingredients were incorporated to provide immediate relief and to treat the cause. For example, sea-ear shell, *Ophiopogon* root, and *Eucommia* bark were used to lower blood pressure and the *Tuckahoe* and mulberry leaves were to act as a sedative. The cicada exuviae were added to treat the tinnitus (Jizong & Zhu 1985).

The cicada species used in Chinese medicine is *Huechys sanguinea*. Its exuviae are also used in prescriptions for migraine headaches and ear infections. The common thread of incorporating cicada skins whenever there was a ringing in the ears or an ear infection probably relates to the cicada's loud singing rather than to any medicinal value. One physiologist told me that cicada exuviae would also quiet babies. Again, the connection was with the cicada's actions rather than any real medical implications. It is the cicada nymphs that are quiet, and by consuming the extract brewed from cicada exuviae, it is thought that the silent stage of the cicada nymphs might influence a baby's loud crying. Myers (1929) noted that early work on the pharmacological properties of dried cicadas was contradictory. Some researchers were able to obtain vesicating effects with a plaster made from ground up cicadas while others were unable to obtain any effects from cicada extracts.

China presents a wide-ranging study in ethnoentomology. Insects, such as the cicadas, are not only used in medicine but have value as religious symbols to the Buddhists, are used as good luck charms, and as symbols of beauty. They were also used as funerary objects (Riegel 1981). Indeed, their practical use in traditional medicine brings a whole new meaning to the term "applied entomology."

Acknowledgment

This project was supported with research grants from the Indiana Academy of Science and the College of Mount St. Joseph.

References Cited

Clausen, L. W. 1954. Insect fact and folklore. MacMillan, New York.

Jizong, S. & C. F. Zhu. 1985. The ABC of traditional Chinese medicine. Hai Feng, Hong Kong,

Myers, J. G. 1929. Insect singers, a natural history, of the cicadas. Routledge & Sons, London.

Richardson, W. N. & T. Stubbs. 1978. Plants, agriculture and human society. Benjamin, Reading, Mass.

Riegel, G. T. 1981. The cicada in Chinese folklore. Melsheimer Entomological Series 30: 15-20.

Used with permission from Bull. Entomol. Soc. Amer. 33: 139-141 (1987).

Insects — An Overlooked Food Resource

Gene R. DeFoliart

As the first 100 years of entomology drew to a close, there is new interest in one dimension of the science that has remained virtually unexplored. Americans and Europeans are learning that throughout human history, insects have played an important direct role in human nutrition. In many cultures, especially in tropical regions, they have served as an important source of animal proteins and fats and of a variety of needed vitamins and minerals.

The traditional and regular dietary use of insects has become less widespread as urbanization and "westernization" have spread. The traditional insects are still important, however, in rural areas where many of the people can but rarely afford to put chicken, goat, fish, or beef on the table. In rural Mexico, for example, Dr. Julieta Ramos-Elorduy of the National Autonomous University in Mexico City finds that insects of more than 100 species are still a regular part of the diet in various parts of the country. As in other countries where insects are used, the most common methods of preparation are roasting or frying; and they may be eaten as a separate dish or mixed with other foods.

These traditional insect foods in Mexico are for sale in the village marketplaces right alongside the same kinds of food products that Westerners are accustomed to buying and eating at home. While the important nutritional contribution of insects is among the low-income population, it should not go unnoticed that some insects are special favorites throughout the country and are eaten by Mexicans of all income levels. They are found on the menus of some of the best restaurants in Mexico City and the other urban centers. One called "gusano blanco de maguey" is the larva of the giant skipper butterfly (*Aegiale hesperiaris*) that develops in the agave, or maguey, plant. Another, "escamol," the immature stages of the ant, *Liometopum apiculatum*, is often sold in tacos. They are served fried with black butter or fried with onions and garlic. A third known as "ahuahutle" or Mexican caviar consists of the eggs of several species of aquatic bugs. The availability of Mexican caviar is now reduced, however, because of contamination of the saline lakes in which the bugs (species of several genera) formerly bred in tremendous numbers.

Mexico is not atypical of many other countries in Africa, Asia, and Latin America where the use of insects as food has been traditional. Insects of 20 or more species are used in many countries and literally hundreds of species on a worldwide basis. The insects are of nutritional importance in the countryside, and same continue to be regarded as "delicacies" in modern urban centers despite Western influence thus providing an additional source of income for rural inhabitants.

In Africa, caterpillars of the giant silk moths (Family Saturniidae) and winged sexuals of the large fungus-gardening termites of the genus *Macrotermes* are particularly popular and their use is widespread. In Zimbabwe (and probably elsewhere), Europeans eat the termites although not in the quantities that are eaten by the local people. In the countries of southern Africa, the "mopanie worm" (caterpillar of the giant silk moth, *Gonimbrasia belina*) is an article of international commerce. The South Africa Bureau of Standards estimated that in 1981, 1600 metric tons were marketed through agricultural cooperatives. This is probably only a fraction of the volume harvested by rural inhabitants. There is a mopanie cannery in the Transvaal, South Africa. In addition, mopanie are eaten by the tons in Zimbabwe and Botswana and exported by the tons from Botswana to Zambia.

Insects are also still widely eaten in Asia, especially in Southeast Asia; and, again, some of the favorites are sold in the markets of the largest cities. One, found in the markets of Bangkok and other urban centers in Thailand, is the giant water bug (*Lethocerus indicus*). W.S. Bristowe wrote in 1932: "It is a great delicacy which is shared by Laos and Siamese alike; it reaches the tables of princes in Bangkok." In the market, the bugs are sold live or steamed; the latter sometimes beautifully arranged in rows. They are prepared for consumption in several ways, one being to fry them in oil; but more commonly they are used as one of the seasonings in a sauce which is served with fried fish. This insect is now exported to the United States where it can be bought in Thai food shops in California. Purchased as whole bugs (known as "mangda"), as bug paste ("nam prik mangda"), or as an alcohol extract ("Mangdana essence"), they are used in the preparation of condiments.

Another insect that is a popular food item in southeastern Asia and could probably find a market in Asian communities in the U.S.A. is the large brown cricket, *Brachytrupes portentosus*. After the wings and legs are removed, the crickets are mixed with ground garlic and salt and then deep-fried until crisp. It is particularly popular in Burma. A former Burmese national now living near Chicago told the author that these crickets are the thing she misses most since coming to the United States. Silkworm pupae (*Bombyx mori*) are another insect item that is widely consumed in Asia and now exported to the United States. In Madison, Wisconsin, they are especially popular in the Korean community.

Fewer insect food items are found in the urban centers of South America. Roasted abdomens of leafcutter ants (*Atta cephalotes* and *A. sexdens*) are sold in Bogota and other Colombian cities, however, and used similarly to popcorn. *Atta* and a wide variety of other insects are used in the South American countryside, and some of them were formerly used in the Caribbean islands. The grubs of two species of beetles, *Rhynchophorus palmarum* (the palm weevil) and *Stenodontes damicornis* are particularly notable for their delicious flavor. Of the former, Sir Robert Schomburgk wrote in 1848 in his *History of Barbados*: "The larva roasted is considered by some of the creoles a great delicacy: it resembles in taste the marrow of beef-bones." Of the latter, Harvard University entomologist, J. C. Bequaert, noted (in 1921) that in earlier times some planters in the West Indies kept blacks whose sole duty it was to go into the woods in search of the larvae. And F. S. Bodenheffer, in his classic, *Insects as Human Food*, stated (1951) that the larva of *S. damicornis* was considered by epicures as "one of the greatest delicacies of the New World." It can be mentioned that the grubs of various species of *Rhynchophorus* are appreciated by indigenous populations in Africa, Asia, and Polynesia, as well as in the Western Hemisphere.

Any general discussion of insects as food would be incomplete without mention of grasshoppers. Grasshoppers and locusts are among the insects used in every country for which entomophagy (eating insects) has been recorded. The plagues that have periodically devastated Africa were greeted with joy by those who had no crops to protect. George W. Stowe wrote in 1905 in his book, *The Native Races of South Africa*, "The arrival of [a swarm of locusts] was hailed by the Bushmen as a glorious time of harvest, as they were esteemed excellent and nourishing food ... The nutritious properties of this food were proved by the fact that during the locust season, the Bushmen increased in flesh, and became rotund and well-conditioned."

Grasshoppers were also the insect food most widely used by native cultures in western North America. They were harvested in abundance and, sun-dried, provided winter provisions. One of the early giants of American entomology, C. V. Riley, was dedicated to the proposition of using the Rocky Mountain locust *Melanoplus spretus*, now considered a migratory form of *Melanoplus sanguinipes*, as food and animal feed during the devastating outbreaks of this species. Riley wrote:

"It had long been a desire with me to test the value of this species (*spretus*) as food, and I did not lose the opportunity to gratify that desire which the recent locust invasions into some of the Mississippi Valley States afforded. I knew well enough that the attempt would provoke to ridicule and mirth, or even disgust, the vast majority of our people, unaccustomed to anything of the sort, and associating with the word insect or 'bug' everything horrid and repulsive. Yet, I was governed by weightier reasons than mere curiosity; for many a family in Kansas and Nebraska was, in 1874, brought to the brink of the grave by sheer lack of food, while the Saint Louis papers reported cases of actual death from starvation in some sections of Missouri, where the insects abounded and ate up every green thing, in the spring of 1875."

Riley was not the first, nor the last, scientist to suggest and in fact urge, that Western cultural biases should be set aside to allow full investigation of the potential of insects as a food resource. In his own time, such other pioneer American entomologists as L. O. Howard and A. S. Packard of the U.S. Department of Agriculture were supportive of his efforts. In more recent times, almost every scientist who has studied the use of insects as food by cultures that have made traditional dietary use of them has called for greater recognition and awareness of their nutritional importance and potential.

In his studies of the Pedi people of South Africa, published by the Witwatersrand University Press in 1959, P. J. Quin concluded that " the recognition and encouragement of their traditional foods and feeding habits could be the means of alleviating, and perhaps even solving, the great problem of malnutrition and disease among these people."

After studying Melanesian food habits in Papua New Guinea, V. B. Myer-Rochow of the University of Waikato in New Zealand concluded in 1973: "If the new Papua and New Guinea government can be persuaded not to accept the European attitude toward insects as human food, it would act to the benefit of vast numbers of natives. Instead of wasting resources in destroying certain insects often regarded as crop pests, the insects themselves should be used. Quite often they represent a higher nutritional value than the vegetable that they have been eating..."

Two researchers at Manipur University in northeastern India, B. Gope and B. Prasad, in 1983 concluded that insects represent the cheapest source of animal protein in the State of Manipur, and recommended that their consumption should be encouraged because many of the people cannot afford fish or animal flesh.

And University of Colorado anthropologist, Darna Dufour, who studied the dietary habits of Tukanoan Indians in southeastern Colombia, states (in 1987) that "the widespread practice of entomophagy warrants further attention in any evaluation of the availability of protein resources" in the Amazonian ecosystem.

Entomologists now believe that insects may come to be regarded as one of the resources with which to meet the nutritional needs of populations that are economically and agriculturally hard pressed. The concept is harmonious with the tenets of low-input, sustainable agriculture. Their persistence in urban markets of the developing world, despite Western acculturation as discussed above, attests to their palatability; and, with a little "push" by appropriate government agencies toward reinstatement of their "respectability," greater use could help yield increased incomes for marginal rural economies. Their traditional and continuing widespread use in the developing world provides a base upon which to rebuild.

Suggested Further Reading

Bodenheimer, F.S. 1951. Insects as Human Food. W. Junk, The Hague, The Netherlands.
Defoliart, G.R. 1989. The human use of insects as food and as animal feed.. Bull. Entomol. Soc. Amer. 35: 22-25.

Chapter 2

Insects Affecting Public Health of Man and Animals

By GARY LARSON

Chronicle Features, 1981

"My goodness, Harold . . . . Now there goes one big mosquito."

Carlos J. Finlay and Yellow Fever: A Discovery

Omelio Sosa, Jr.

Carlos J. Finlay was born in Camaguey, Cuba, Dec. 3, 1833, to Edward Finlay, a Scottish physician, and Eliza de Barres, a Frenchwoman born in Trinidad. He graduated in 1855 with an M.D. from Jefferson Medical College in Philadelphia, where he was noted as persistent and patient. Sila Weir Mitchell, Finlay's classmate, tried to persuade Finlay to practice in the United States, but Finlay decided to return to Cuba. In 1860 he traveled to France to further his studies, returning in 1864 to Matanzas, Cuba, and later establishing his residence in Havana in 1865.

In 1867, Havana experienced an outbreak of cholera. Finlay, following his methods of observation and investigation, concentrated his attention in the so-called *Zanjia Real* (Royal Ditch). He noticed that a great number of cholera patients lived in areas through which the contaminated waters of the Royal Ditch ran. On June 27, 1868, Finlay wrote an extensive letter to the editor of *Diario de la Alarina*, a Havana newspaper, in which he presented his observations and outlined sanitary procedures that should be adopted (Rivas-Aguero 1983). His letter went unpublished because the censor considered it a criticism of the authorities and an indictment of their failure to combat the epidemic. Consequently, many lives were lost unnecessarily. It took almost 20 years for the government to carry out the same measures Finlay had proposed. A similar fate was to plague Finlay's research on yellow fever.

In 1879, the United States sent to Cuba a medical commission led by S. E. Chaille to investigate methods of controlling yellow fever; Finlay was appointed adviser to the commission by the Spanish Governor of Cuba. The commission failed, but the experience inspired Finlay to explore the topic. At a meeting of the International Sanitary Conference in Washington, D.C., in 1881, Finlay presented for the first time his views about the transmission of yellow fever. He listed three conditions necessary for the spread of the disease: One person must have the disease; there must be a second, susceptible person; and there must be a vector to transmit the disease from the sick person to the healthy one (Finlay 1881a, del Regato 1968, Cepero 1977). This medical concept was completely new.

Finlay raised the possibility that with the help of experimentally infected mosquitoes, mild but immunizing cases of yellow fever could be produced (Finlay 1898). Upon his return to Havana, he met with the captain general of Spanish troops, and 20 Spanish soldiers volunteered for the experiment. Volunteers had to be recent arrivals who were not acclimated. Finlay believed this was necessary because yellow fever had been endemic in Cuba for more than two centuries, and a great number of the inhabitants were immune to the disease.

The volunteers were quartered on the heights of the Cabana across the bay to minimize their exposure to mosquitoes. Finlay inoculated five volunteers and kept detailed medical records (Finlay 1886). Two soldiers were diagnosed by the attending physician at the military hospital as having "abortive yellow fever," and one was found to have "regular yellow fever." The fourth soldier suffered only from continued headache and a slight fever (38.2°C) on the fifteenth day after inoculation. The fifth soldier did not return to Finlay's office. None of the 15 uninoculated soldiers contracted yellow fever during the period of Finlay's tests from June 28 to September 1881.

Finlay (1881b) presented the results of these tests to the Royal Academy of Medical Sciences of Havana in August 1881. His conclusion was that the *Culex* mosquito was the only agent-transmitter of the disease. *Culex* was later renamed *Stegomyia fasciata*, and today is known as *Aedes aegypti*. Unfortunately, no one showed any interest in his investigations. His presentation served only to label him sarcastically as the Mosquito Doctor. His paper (Finlay 1881b) was translated and published by Rudolph Matas (Finlay 1882) in the *New Orleans Medical and Surgical Journal*.

Encouraged by the fact that a single bite had not caused a fatal yellow fever attack, Finlay continued his human experiments. He found a new source of volunteers among unacclimated Jesuit priests housed at San José, a farm outside Havana, where there had not been a case of yellow fever in a decade. Finlay (1886) gave a detailed medical account of the fate of inoculated volunteers. Case 71 was Father Gutierrez-Lanza, and it was to him that Finlay confided how he came to suspect the mosquito as the possible vector (del Regato 1968). One night, as Finlay sat beside his bed reciting the rosary, he was being disturbed by a persistent mosquito. It was then he first conceived what had not occurred to any man. The mosquito was the vector of yellow fever.

Finlay, went on to accumulate 104 records of inoculations; most volunteers developed abortive cases of yellow fever. Only four developed yellow fever in later years, two of them fatally. Furthermore, during the years between the discovery and its probe by the medical commission led by Walter Reed in 1900, Finlay wrote 42 papers on the subject, published mostly in British and American medical journals.

Three more medical commissions were sent to Cuba. George Sternberg, leader of the second commission, in April 1888, did not believe Finlay's theory and consequently did not consider it. After the Spanish American War, the United States sent a third commission, led by E. Wasdin and H. D. Gedding, which also failed to find a method of controlling yellow fever. A fourth commission, led by Walter Reed also almost failed; Reed believed Finlay's finding lacked a scientific base.

On Aug. 1, 1900, members of the Reed Commission visited Finlay at his home. Finlay gave Reed all the information he had accumulated in 19 years of work on yellow fever. He also gave Reed mosquito eggs and explained that they were the eggs of the mosquito that transmits yellow fever (Rivas-aguero 1983). The next day Reed returned to the United States, leaving behind James Carrol, Jesse Lazear, and Aristides Agramonte. Carrol allowed himself to be bitten and six days later was stricken, although not fatally, with yellow fever. Private William H. Dean of the Seventh Cavalry, a volunteer, also developed yellow fever within six days. On Sept. 13 Lazear allowed an infected mosquito to bite him. He died Sept. 25, 1900. This unselfish act focused attention on the mosquito theory and accelerated its solution.

Reed went to Havana and then to Indianapolis after reading Lazear's notebook. There, on Oct. 23, 1900, Reed presented a paper, "The Etiology of Yellow Fever. A Preliminary Note," to the American Public Health Association. He discussed the cases of Carrol and Lazear, concluding (as Finlay had done in 1881) that the mosquito "serves as the intermediate host for the parasite of yellow fever" (del Regato 1968, Cepero 1977, Rivas-Aguero 1983). Reed also stated, "We here desire to express our sincere thanks to Finlay, who accorded us a most courteous interview and has gladly placed at our disposal his several publications relating to yellow fever during the past 19 years; and also for ova of the variety of mosquito with which he had made his several inoculations. With the mosquitos thus obtained, we have been able to conduct our experiments. Specimens of this mosquito forwarded to Mr. L. O. Howard, Entomologist, Department of Agriculture, Washington, D.C., were kindly identified as *Culex fasciatus* Fabr. The mosquito serves as an intermediate host of yellow fever, and it is highly probable that the disease is only propagated through the bite of this insect" (Reed, *et al.* 1900).

The site for further experiments by the commission was appropriately named Camp Lazear. It was located near San José farm, where Finlay had conducted inoculation experiments with young Jesuit priests. Two 4- by 6-meter buildings were constructed and placed 73 meters apart. The first building was built with inadequate ventilation but screened to keep mosquitoes out. Bedding and clothing soiled with vomit, urine, and feces of yellow fever patients were brought in. The second building was well ventilated, divided by a metal screen, and provided with two separate entrances. On Nov. 20, Privates R. P Cooke, L. E. Folk, and W G. Jernegan entered the first house. Private J. R. Kissinger was inoculated with infected mosquitoes and entered the second building on one side of the screen, and clerk J. J. Moran, who was not inoculated, entered on the other side of the screen.

By Dec. 8, Kissinger had contracted yellow fever as certified by four doctors, Gorgas, Finlay, Guiteras, and Albertini. All other volunteers were healthy. On that day, Agramonte inoculated three unacclimated Spanish volunteers: A. Benigno, B. Precedo, and N. Fernandez. By Dec. 13 all three had contracted yellow fever (Reed *et al.* 1901, del Regato 1968). These experiments served to confirm Finlay's results from similar experiments. Finally the mystery was solved and, more important, believed. Evidently, the world had not been ready to accept or even listen to what Finlay had been saying. Those who listened were so entrenched in their own thinking that for whatever reason, did not open their minds to Finlay's discovery. Uncounted lives could have been saved if scientists had heeded what Finlay had been proclaiming for years: that *Aedes aegypti* was the yellow fever vector.

In September 1894, at the Eighth international Congress of Hygiene and Demography in Budapest, Hungary, Finlay presented the fundamentals of a program to eradicate yellow fever: isolation of patients, fumigation and screening of houses, and extermination of breeding grounds of mosquitoes. Following these basic measures, the battle was waged against the mosquito by W. C. Gorgas, who eradicated the mosquito from Cuba and later from Panama. The control of yellow fever was a major factor in the success of the United States in completing the Panama Canal. The French had previously tried to build the canal but had failed, mainly because thousands of their workers had succumbed to yellow fever.

For years Finlay's theory was ignored; the results of his experiments failed to convince the incredulous. It was not until the Reed Commission confirmed Finlay's work that it gained wide acceptance. Possibly because of this, Reed's name has been permanently linked and credited with the discovery of the mosquito as the vector of yellow fever. Nevertheless, two international Congresses on the History of Medicine, one in Madrid in 1935, and one in Rome–Palermo, Italy, in 1954, unanimously proclaimed that "To him [Finlay] and only him, should be credited the discovery of the vector of yellow fever" (Rivas-Aguero 1983). The latter congress, in addition to ratifying the statement of the 1935 congress, recommended a "campaign to disseminate this information and to correct erroneous textbooks." However, very little has been done in this respect. Most textbooks mention only in passing, if at all, Finlay's contribution. This is a sad and unjust omission that perpetuates even today.

Finlay's discovery unlocked the mystery that had puzzled scientists about this dreaded disease. This knowledge enabled others to eradicate yellow fever from the tropics. Finlay went on to be appointed as the first Sanitary Director of the Republic of Cuba when it became independent May 20, 1902. In recognition, he received an honorary degree from Jefferson Medical College in Pennsylvania, the Mary Kingsley Medal from the Institute of Tropical Medicine in Liverpool, and the French Legion of Honor. He was nominated for a Nobel Prize by Donald Ross in 1905, by John W. Ross in 1907, and by the Academy of Medicine of Havana in 1912. It is likely that the numerous contemporary claims kept the members of the academy from rendering him justice. Finlay retired after seven years of service to the young republic, and died in Havana in 1915 (del Regato 1971, 1987). The centennial of his birthday, Dec. 3, 1833, was dedicated in his memory by Congress of the Pan American Association of Medicine as the Americas Day of Medicine, and the U.S. Congress dedicated a special session to his memory (Cepero 1977). Also, the subgenus *Finlayia* of the genus *Aedes* is named in Finlay's honor. It is only fair to give Finlay the credit he deserves for his immense contribution to mankind. In so doing, this acknowledgment does not diminish contributions made by others.

References Cited

Cepero, G. R. 1977 Notes on the evolution of medicine in colonial Cuba. J. Fla. Med. Assoc. 64: 570-577.

del Regato, J. A. 1968. Carlos Finlay and the carrier of death. Americas; Magazine of the Panamerican Union. May.

del Regato, J. A. 1971. Carlos Finlay and the carrier of death. The cycle of successful scientific discovery. Alumni Bulletin,Jefferson Medical College, summer: 1-16.

del Regato, J. A. 1987 Carlos Finlay and the Nobel Prize in physiology or medicine. The Pharos of Alpha Omega Alpha. 5: 5-9.

Finlay, C. 1881a. pp. 119-120. In Proceedings of the International Sanitary Conference, Washington, D.C., Protocol 7 Government Printing Office, Washington, D.C.

Finlay, C. 1881b. El mosquito hipotéticamente considerado como el agente de transmisión de la fiebre amarilla. An. Acad. Cienc. Méd. Fis. Natur. (Habana) 18: 147-169.

Finlay, C. 1882. The mosquito hipothetically [sic] considered as the agent transmitter of yellow fever poison. New Orleans Med. Surg. J. 9: 601-616.

Finlay, C. 1886. Yellow fever: its transmission by means of the *Culex* mosquito. Amer.J. Med. Sci. 92: 395-409.

Finlay, C. 1898. A plausible method of vaccination against yellow fever. Phila. Med.J. 1: 1123-1124.

Rivas-Aguero, M. 1983. Historia de un grande de la patria: Carlos Finlay. El Camaguevano Nov.-Dec.: 19-27.

Reed, W, J. Carrol, A. Agramonte & J. W Lazear. 1900. The etiology of yellow fever: a preliminary note. Proceedings of the 28th annual meeting of the American Public Health Association, Phila. Med. J. 6: 790-796.

Reed, W, J. Carrol & A. Agramonte. 1901. The etiology of yellow fever: an additional note. J. Am. Med. Assoc. 36: 431-440.

Used with permission from Bull. Entomol. Soc. Amer. 35: 23-25 (1989).

Malaria Control and Eradication in the United States

Harry D. Pratt

Malaria has been one of the major diseases afflicting mankind for centuries. Prior to World War II malaria was so common in the United States that many people kept the malaria drug, quinine, on their dining tables and took the medication with their meals. In the first part of the twentieth century Gorgas and Leprince conducted intensive control operations against the mosquitoes which transmitted malaria and yellow fever, making possible the building of the Panama Canal. During the Great Depression of the 1930s (see graph, p. 57) over 100,000 cases of malaria were reported each year in the United States, but the number of cases was probably grossly underreported. Dr. L. L. Williams, the U. S. Public Health Service expert on malaria, estimated that as many as 3 to 4 million cases of malaria actually occurred each year. During the 1930's the U. S. Public Health Service, the Tennessee Valley Authority, and various state health departments demonstrated that area-wide cooperative programs directed by a physician-entomologist-engineer team could control malaria in parts of this country. By 1942 the number of cases of malaria officially reported dropped to about 60,000.

When World War II began, the U. S. Public Health Service followed this same physician-entomologist-engineer team concept and started the Malaria Control in War Areas (MCWA) program with headquarters in Atlanta, Ga. under the overall direction of Dr. L. L. Williams and Mr. Mark D. Hollis, with Dr. T. H. Stubbs as Chief of the Medical Division, Dr. G. H. Bradley as Chief of the Entomology Division, and Mr. N. Rector as Chief of the Engineering Division. The MCWA program was carried out in areas contiguous to military reservations and major war production installations and surpassed in magnitude anything previously known. During the period of maximal military training and industrial production, the MCWA control operations were conducted at approximately 2200 localities in 15 southeastern states, California, and parts of the Caribbean.

A fundamental decision was made at the beginning of the MCWA program that operations would be focused on the specific vectors of malaria: *Anopheles quadrimaculatus* east of the Rockies,

MALARIA- Reported Cases by Year, United States
1930-1988

NUMBER OF CASES

YEAR

TVA Malaria Control Program

WPA Malaria Control Program

War Areas Control Program

Relapses-Overseas Cases

Malaria Eradication Program

Relapses From Korean Veterans

PRIMAQUINE Treatment of Servicemen
Returning From Malarious Areas

Returning Vietnam Veterans

Asian Refugees

Small epidemics
in California

Anopheles freeborni west of the Rockies, and *Anopheles albimanus* in the Caribbean. Such selective control measures were made possible by utilizing entomological survey and inspection services to the fullest extent. This procedure reduced the total costs of the MCWA program by an estimated 15 million dollars during World War II.

At first the program stressed drainage and filling of malaria mosquito larval breeding places or larviciding with oil or Paris green, a copperaceto-arsenite insecticide. But DDT became available for civilian programs in 1945. So this new "wonder insecticide" was used both as a larvicide and as a residual spray applied to the walls of homes where for months it would kill infected mosquitoes before they could transmit the malarial parasite to uninfected people, thus breaking the chain of malaria transmission.

Despite all this work in the United States, there was an increase in the number of cases of malaria officially reported at the end of World War II, 62,763 cases in 1945, probably because veterans returning home from the China-Burma-India and Mediterranean theaters of war had acquired their infection overseas.

When World War II ended, the personnel of the MCWA program were transferred to the National Malaria Eradication Program of a new U. S. Public Health organization, the Communicable Disease Center (CDC), officially beginning July 1, 1946. This cooperative program of USPRS and many State Health Departments was based largely upon residual DDT spraying of homes to kill infected malaria mosquitoes. From 1945 to 1952 some 6,500,000 sprayings were made (1,365,000 homes in 1948). Many other factors were involved in the decreased number of malaria cases reported in this country. The reporting system was changed from the simple notice of an apparent clinical case. The various state health departments reported malaria cases only after microscope slide identification of the malaria parasite in the patient's blood. Many people migrated from the malarious South to better jobs in the non-malarious, industrial North. Homes were screened. Pyrethrum aerosol sprayers became household items in mosquito-infested areas. New drugs such as atabrine, chloroquine, and primaquine replaced the traditional quinine. The number of cases of malaria dropped dramatically from over 60,000 in 1945 to 2,184 in 1950. Eradication operations were stopped in 6 states in 1950, in 7 states in 1951, and in all states in 1952 when the Eisenhower administration began major governmental economies.

At the same time that the National Malaria Eradication Program was closing down, the United States became deeply involved in the Korean War. As veterans returned from Korea, many of them came down with malaria. The number of cases increased to 5600 in 1951 and 7023 in 1952, of which approximately 50 were contracted in the United States. Almost all of the Korean cases were of the *Plasmodium vivax* type of malaria, which tend to have relapses after the primary attack. So these patients were treated with chloroquine and primaquine to prevent relapses of malaria. As the number of veterans returning from Korea decreased, the number of cases of malaria continued to drop each year — from 1310 in 1953 to 132 in 1957. From 1957 to 1965 each year less than 200 cases of malaria were reported in the United States, almost all of them acquired overseas never with more than 3 cases in any one year actually due to mosquito transmission in this country. Malaria eradication was essentially achieved at this time.

Beginning in 1965 the United States became increasingly involved in war in Vietnam. As veterans returned to this country, the number of cases with onset in this country increased to 678, some 563 of them in military personnel. This number grew larger for several years and in 1970 reached a peak of 4247 cases, 4096 of them in military personnel. While almost all of the Korean War veteran malaria cases were of the *Plasmodium vivax* type typically found in temperate zone countries, many of the veterans from tropical and subtropical Vietnam were infected with the *Plasmodium falciparum* type of malaria often found in tropical countries, while others had the *Plasmodium vivax* type.

Since the end of the Vietnam War, the number of cases reported in military personnel has decreased greatly, varying from 41 in 1973 to 5 in 1976. On the other hand there has been a considerable increase in the number of cases in foreign visitors or immigrants and in U.S.A. civilians. Most of these U.S.A. cases were contracted in Asia, Africa, or Central or South America in such people as Peace Corps workers, missionaries, safari travelers, or members of tour groups. Beginning in 1980, the number of cases of malaria with onset of illness in the United States has averaged about 1000 cases each year with only a few resulting from mosquito transmission in this country.

Following President Carter's decision to admit up to 14,000 Asian refugees a month in 1979, the number of malaria cases increased to 1864 in 1980, due in part to infected Indochinese refugees. In 1986, an outbreak of 28 cases was reported in southern California near Carlsbad, San Diego County and five additional ones in northern California. In 1988, an outbreak of 30 cases was confirmed in San Diego County near Lake Hodges. All but four cases were Hispanic migrant farm workers, the other cases were local residents living near the site of the outbreaks but with no previous history of malaria. While the malaria cases in northern California may have been due to transmission by *Anopheles freeborni*, the vector in the two outbreaks in San Diego County in 1986 and 1988 probably was the recently described malaria mosquito, *Anopheles hermsi*.

Despite the widespread presence of malaria mosquitoes through-out the United States, a susceptible human population, and the importation of thousands of cases of malaria acquired overseas, only 16 episodes of introduced malaria have been reported between 1950 and 1988, i.e., malaria due to transmission from an infected individual coming into the United States. In each of these episodes the patients were given prompt medical treatment and careful surveys were made to locate and/or control the malaria mosquitoes. There-fore, surveillance continues to the present time, particularly by health departments, mosquito abatement districts, and the military.

Acknowledgement

I appreciate very much C. S. Stanley and the Centers for Disease Control for providing the graph on Malaria – Reported Cases by Year, United States. 1930-1988.

Suggested Further Reading

Andrews, J. M., J. S. Grant, and R. F. Fritz. 1954. Effects of suspend-ed spraying and of imported malaria in the U. S. A. *World Health Organization Bull.* 11: 839-848.

Barr, A. R. and P. Guptavanij. 1989. *Anopheles hermsi* n. sp., an unrecognized species of the *Anopheles maculipennis* group. (Diptera: Culicidae). *Mosquito Systematics* 20 (3): 352-356.

Boyd, M. F. (ed.) 1949. *Malariology*, Vols. I and II. W. B. Saunders Co., Philadelphia. 1643 pp.

Bruce-Chwatt, L. J. 1980. *Essential Malariology*. W. Heinemann Medical Books Ltd., London, 354 pp.

Centers for Disease Control. 1962-to date. Malaria surveillance annual summary. USDHHS, PHS, CDC, Atlanta, Ga.

Russell, P. F., L. S. West, R. D. Manwell, and G. MacDonald. 1963. *Practical Malariology*. Oxford University Press, London. 750 pp.

100 Years of Entomology in the Department of Defense

Harvey A. Shultz

Introduction

On July 21, 1988 a group of 50 pest management professionals, most of them entomologists employed by the Department of Defense (DoD), purposefully filed into a conference room in Washington, DC. Commissioned officers and civilians alike, representing the Army, Navy and Air Force, and other federal agencies, they came to coordinate entomology, pest management and related functions at the 128th meeting of the Armed Forces Pest Management Board (AFPMB). The pace of the formal proceedings was intense, and the agenda ambitious. A dozen standing and *ad hoc* committees made recommendations regarding pesticides, dispersal equipment, training and certification of pest controllers, repellents, stored products, pest control research, and much more. Debate was spirited, but inevitably the voting members moved towards a consensus. Sitting in was Mr. William A. Parker, III, the Deputy Assistant Secretary of Defense for Environment to whom the Executive Director of the AFPMB reports. Mr. Parker used the occasion to praise the Board for its field-oriented professionalism and environmental awareness. Appropriately, the meeting took place at Walter Reed Army Medical Center, named after the Army Medical Corps officer whose work on the transmission of yellow fever by mosquitoes almost a century ago marked the beginning of modern military entomology.

In 1889 there was no organized entomological resource available to military commanders. As each challenge presented itself, professionals, often physicians like Major Reed, were pressed into service. What is so special about entomology in today's military that demands the level of attention described above? Let's answer this question by tracing military entomology from its modest beginnings to present times.

The Early Days

One hundred years ago, the role of insects in disease transmission was poorly understood. Little had changed since the 1870's when the Surgeon General of the Navy quoted a prevailing theory of the day, "that yellow fever poison propagates itself rapidly and spreads by creeping along the ground . . ." Mosquitoes, flies, lice, ticks, and the like were regarded mostly as nuisances. Accordingly, there was no incentive for the military to engage the services of scientists knowledgeable in insect matters. In the 1880's and 1890's a few medical officers contributed observations, mostly concerning the enormity of mosquito swarms at Army camps and the primitive measures used to gain relief.

Then in 1898, a poorly prepared United States went to war with Spain. There were only 369 American battle casualties during the 118-day Spanish-American war. By contrast, 1,939 men died of disease, primarily typhoid fever contracted in volunteer camps in the United States. A Board of Army Medical Officers including MAJ Walter Reed, concluded that the spread of the disease was due primarily to contact between soldiers and flies. (Things might have been worse if the war had lasted longer since the Surgeon General's share of the war budget was a meager $25,000!)

Surprising to some today, but not to military entomologists, these figures were predictable because casualties due to disease, primarily arthropod-borne disease, have exceeded battlefield casualties in *every* war fought by the United States!

Upon the conclusion of hostilities, MAJ William Crawford Gorgas, began the "sanitation" of Havana. However, an on-going outbreak of yellow fever continued unabated until 1901, when Walter Reed and associates of the Yellow Fever Commission provided evidence that the *Aedes aegypti* mosquitoes could transmit yellow fever from man to man. Within-eight months yellow fever was eradicated from Havana.

Gorgas put his Havana experiences to the test in Panama over the objections of some groups which did not agree with the mosquito transmission "theory." Fortunately, President Roosevelt had good entomological instincts and backed the Army Medical Department. The malaria rate was reduced from 1263/1000/annum (many individuals suffered multiple cases each year) in 1906 to 76/1000/-annum in 1913. The United States might have failed to complete the Panama Canal, as did the French earlier, were it not for Gorgas' three part program – screening all dwellings, fumigating dwellings of victims, and eliminating or oiling standing water. By his own estimate over 70,000 lives were saved – primarily from yellow fever and malaria.

By the time WWI broke out, the link between microbial life, disease, and vectors was well established, but pieces of the puzzle were still missing. Malaria was not a serious problem among U.S. troops in Europe during the war, but almost 10,000 cases occurred in training camps in the southern U.S.A. In Macedonia, however, malaria-weakened British and German armies were unable to advance for three years.

WWI was fought largely in trenches, and as a consequence trench fever, a new disease known to be transmitted by body lice, was widespread. In response, the British Expeditionary Forces added two entomologists to each of their sanitary units. Delousing stations were established in Europe, and conditions were much improved by the time U.S. troops entered the conflict.

The British also cooperated with the U.S. Army Surgeon General in regard to experimental work on louse control and other vector problems. Poisonous gas, so deadly in the trenches of Europe, was tested for control of insects and diseases. Dry cleaning processes which controlled body lice on garments were perfected for the Quartermaster Corps. High frequency sound, not surprisingly, failed to do the same. More work was done in properly sizing mosquito meshes to exclude mosquitoes from cantonment buildings. For the first time a handful, no more than 8, entomologists were commissioned in the Army, while others served as non-commissioned officers assisting in the control of insects in Army camps.

The War Department called on the United States Department of Agriculture (USDA) Bureau of Entomology for assistance in controlling grain pests infesting immense warehouses in Brooklyn, NY and to protect forest resources needed to manufacture oars, handles, and airplanes – yes, airplanes!

References to military entomology in the literature between the world wars is sparse. However, an intriguing character, Clara Ludow, performed mosquito taxonomy for the Army on a voluntary basis from 1901 till her death in 1924. Ms. Ludow, the sickly daughter of a civil war medical officer, developed her interest while visiting her brother, an artillery officer stationed in the Philippine Islands. Her fascination with mosquitoes lead to a Ph.D. at age 55 and a final resting place in Arlington National Cemetery. Some consider her to have been the first "military entomologist," as among her publications are many clearly directed towards military problems.

WWII

For the past 45 years DoD has been the largest employer of medical entomologists in the world. But the military learned the importance of entomology the hard way – between the world wars only 14 Army entomologists were commissioned in the reserves! This is especially ironic, since during the same period knowledge of vector-borne disease grew impressively.

It seems predictable today that a little dapple-winged mosquito, the *Anopheles*, would cause five times more casualties among U.S.A. troops during WWII than would battlefield injuries! "Malaria discipline" had not been developed and the prevailing attitude, as expressed by a high ranking officer on Guadalcanal was, "We are out here to fight Japs and to hell with mosquitoes." The combination of effective vectors, susceptible troops, infected native and Japanese reservoirs, and prolific natural and battle-caused mosquito breeding sites set the stage for what was about to happen. Although only 1000 American lives were lost to arthropod-borne disease in WWII, the toll in lost manpower was staggering – almost 2 million cases totalling 24 million man-days lost! In May of 1943, General MacArthur observed that, "This will be a long war, if for every division I have fighting the enemy, I must count on a second division in the hospital with malaria and a third division convalescing from this debilitating disease." Recognizing the impact of malaria on morale and fighting ability, a history of the U.S. Army medical department noted that, "The Pacific and Asiatic campaigns would have been impossible without control of malaria." Victory was possible because in 1940, the Army Surgeon General realized the U.S.A. could be drawn into a world conflict, and began to organize the Medical Department to meet any contingency. The new Preventive Medicine Service included a Malaria Control Branch and (under another Service) an Insect and Rodent Control Branch.

In 1941 the Naval Medical Department established the Hospital Volunteer Specialist Group and, the first two Navy entomologists (LTJG William K. Lawler and LT Paul Woke) were commissioned. By the end of WWII they had been joined by over 200 colleagues. In 1942, U.S. troops experienced a malaria rate on the Island of Efate, New Hebrides of 2600/1,000/annum! A team including Ensign Kenneth L. Knight, the first Navy entomologist to work in a combat zone, reduced the rate to essentially nil. Due to the devastating effect of malaria on U.S. troops in the Pacific in 1942, entomologists were recruited from USDA, the United States Public Health Service, and universities to staff survey and control teams. The South Pacific Malaria and Insect Control Organization was a joint Army-navy-allied group. By 1944 the War and Navy Departments had put 771 specially trained personnel in the field. Army entomologists staffed 17 malaria survey detachments; still other units were involved in control. The Navy Division of Preventive Medicine established epidemiology units, which by 1944 numbered 122. Over 100 malaria control units led by entomologists and medical officers were distributed throughout Naval units and in forward areas. Navy entomologists also served in China, North Africa, the Caribbean, and Central Africa. William B. Herms, of the University of California, who had served as an officer in the Sanitary Corps of the Army in WWI, returned to duty for WWII to join 45 entomologists he helped train at Berkeley's Division of Entomology and Parasitology. Malaria cases overseas peaked in 1943 and then dropped dramatically. Due to the concurrent arrival of trained personnel and DDT, overall rates in the South Pacific dropped from 208/1000/annum in 1943 to 5/1000/annum for the first half of 1945. As a spin-off benefit, the entomological requirements of the day did wonders for the worldwide taxonomy of *Anopheles*, as keys were developed to aid in control efforts.

Malaria, however, was not the only arthropod-borne disease to adversely affect U.S. troops. During a 3-month period in 1944, 1000 of 3500 U.S. troops assigned to Saipan contracted "break-bone fever," or dengue. Scrub typhus, a debilitating disease transmitted by chigger mites, accounted for 18,000 casualties among allied troops in the Pacific-asiatic theater. There, mortality from scrub typhus was considerably higher than from malaria. In New Guinea elements of the 6th Infantry Division were rendered non-effective after suffering 931 cases and 34 deaths within a few weeks of hitting the beach. Air Force personnel in Australia suffered an overall rate of 750/1000/annum. Control was accomplished by eliminating the coarse grass harboring wild rodents (the natural disease reservoir), and treating uniforms with the recently developed M-1960 clothing repellent.

Q fever, a rickettsial disease barely known at the time Pearl Harbor was attacked, may have accounted for 75% of the cases diagnosed as atypical pneumonia in northern Italy. This tick-borne disease decreased unit strength by as much as 30% and later caused great concern when 400 cases occurred in troops returning to Virginia.

Early in the war USDA was asked to coordinate disease prevention and vector control research projects. Of initial concern were the development of new repellents and pesticides to control scabies and the human body louse. The war stimulated a level of research in the control of arthropod-borne diseases unprecedented in history at the USDA lab in Orlando, Florida.

The history of technical developments in pest control resulting from wartime needs included the initial field use of a synthetic organic pesticide (DDT) for vector control, the development of insecticide aerosols, the development of effective vector repellents and advances in fumigation.

The discovery in 1943 that DDT, applied as a residual spray on surfaces, was lethal to adult mosquitoes, led to a feasible plan for malaria eradication by interruption of the transmission cycle. (This strategy would later be validated by the World Health Organization in its global eradication effort.)

Later in the same year, louse-borne typhus broke out in allied-occupied Naples. Recalling the loss of life due to typhus at the end of WWI, the military dusted 2.5 million people in Italy (and later

2 million more along the Rhine River) with DDT. As a result, no new cases of typhus were reported. Concerns as to environmental implications and other potential adverse effects would come later, but world-wide dependency on residual synthetic, organic pesticides had begun.

The need for a simple way to disperse insecticides in the field had become critical. Malaria cases exceeded battle casualties in the Sicilian campaign. At West African airbases the malaria case rate surpassed 2000/1000/annum. Less heralded than the development of the atomic bomb was a quest (marked by similar urgency) for another bomb – the bug bomb. Two USDA researchers with support from the Army Medical Center, Walter Reed Hospital and the Aero-medical Laboratory, Wright Field, Dayton, Ohio produced the prototype, a 5 pound cylinder containing the newly available Freon 12, pyrethrin, and sesame oil. The aerosol was produced by inverting the cylinder and opening the valve. Within months American industry was able to improve the container and nozzle and deliver supplies to units in the field, a feat which must be considered amazing by contemporary standards. The aerosol insecticide bomb went on to launch a major postwar industry.

The Air Force's most noteworthy contribution to military entomology, appropriately, has been in aerial dispersal of insecticides. But the need for aerial spray preceded the separation of the Air Force from the Army. In cooperation with the USDA Orlando laboratory, the Army Air Force (AAF), in 1943, began developing apparatus for aerial dispersal of DDT to control adult and larval mosquitoes. The Fifth Air Force, equipped with L-4 aircraft, first used the new technique in New Guinea in 1944. The operation was directed by Captain W.C. McDuffie who later became chief of the USDA Laboratory for Insects Affecting Man and Animals. The original 32-gallon insecticide tanks, venturi pumps and pipe-like dispersal systems, were soon replaced with 625 gallon tanks, hydraulic pumps and multi-nozzle booms mounted below the wings of B-25s and C-47s. Some Pacific beachheads were sprayed hours before troop arrivals. In the China-Burma-India theater, the Philippines, and the Mariannas, volunteer aerial spray units were also formed; but for lack of entomological input, success varied.

The Army and Navy have also developed and deployed aerial spray systems with emphasis on supporting forward-deployed troops, in relatively small areas. In 1943 two Navy entomologists "jury rigged" a French Amiot bomber in Morocco to disperse Paris green for malaria control. The following year, one of these officers, LT Joseph Yuill, adapted a system onto a Coast Guard HM5-4 in New York City, and insecticides were dispersed for the first time from a helicopter. The Army and Navy have continued to progress with helicopter-mounted aerial spray technology, while leaving large area, fixed-wing applications to the Air Force.

In 1946 a special Aerial Spray Flight (SASF) was created at Greenville Army Air Base in South Carolina to continue research and to provide services to military installations in the United States. Today's Air Force aerial spray capability is part of the 356th Tactical Airlift Squadron at Rickenbacker Air National Guard Base, Ohio. Large, but agile, C-130 aircraft are equipped with a modular spray system capable of high, low and ultra-low volume insecticide delivery.

After WWII few entomologists remained on duty, and most of the epidemiology units were disbanded. An exception was the Malaria and Mosquito Control Unit at Naval Air Station (NAS) Banana River, FL which was moved to NAS Jacksonville, FL in 1947 and eventually designated the Disease Vector Ecology and Control Center (DVECC). In 1959 a sister unit was established in Alameda, CA.

In 1949 DVECC Jacksonville began the development of insecticide dispersal equipment with the initial work on thermal aerosol (fog) applications. These dispersal units, which evolved from DVECC research, helped produce the cold foggers of the 1960s, the ultra low volume (ULV) units of the 1970s, and in 1978 computer analysis of ULV droplets.

In 1951, this same unit was also the first to conduct training for civilian and military pest controllers. This type of training was made mandatory for commercial applicators by the Environmental Protection Agency (EPA) almost 25 years later. In the early 1970s the four week version of the training taught at both DVECCS and at Army and Air Force schools was applauded by EPA as a model that the states might well follow.

Korea and Vietnam

Military entomologists also played a significant role during the Korean conflict, where arthropod-borne diseases again threatened military operations.

In 1950 modified C-40s, later L-5s on loan from the Army, and finally T-6 aircraft were deployed to protect United Nation forces in Korea. The Korean spray unit, trained by SASF experts, used 20% DDT oil solution at 12 sites to control mosquitoes which were responsible for almost 1500 cases of malaria and Japanese B encephalitis in the first half of 1950. Later, as dysentery cases increased, Pusan, Taegu, and Seoul were sprayed to control flies.

On the ground, relapsing fever and typhus were also prevalent. One hundred per cent of the internees at the Koje D Prisoner of War Camp became infested with DDT-resistent lice and innovative measures were required. Mass power delousing with the newly

Rufus H. Vincent, Air Force Director of Installations helped pioneer this concept.

The Army now employs 52 civilian entomologists, the Navy 21 and the Air Force 8, many of whom provide command-wide or geographical integrated pest management support. They implement programs which insure that insects and other living things do not interfere with operations, destroy property or material, or adversely affect human health. These pest management professionals develop long range, comprehensive, environmentally compatible, economical pest management plans for supported installations. Plans are updated and reviewed periodically on-site for compliance. Of special interest is the safe selection, storage, formulation, application, and disposal of pesticides, though non-chemical means of control are preferred wherever feasible. Civilian entomologists also participate in training pest controllers and pest control contract inspectors. Entomologists write and/or tailor contract specifications for pest control work to help insure receipt of quality pest control services. Where in-house work prevails, the wooden pest control shops of the 1950s have been replaced with well engineered, state-of-the-art facilities designed to protect workers and the environment from the adverse effects of pesticides. Mission-oriented engineer entomologists, by virtue of their responsibilities often become "generalists." They grow professionally on the job to develop expertise in such diverse fields as animal damage control (in some cases even including bird/aircraft strike hazard reduction), weed control, and wood protection.

The entomology program in the Services today is characterized by the readiness and quality of its professional personnel. Most DoD civilian entomologists hold advanced degrees and are affiliated with the American Registry of Professional Entomologists (ARPE). Active duty entomologists today number 72 in the Army, 39 in the Navy, and 16 in the Air Force, while many more entomologists serve in the Reserve and National Guard. The masters degree is prerequisite for commissioning and most officers hold Ph.Ds. ARPE registration is the norm. Senior officers fill top level environmental and medical staff positions, but none forget that they are field troops first. Among the 241 killed when the Marine barracks in Beirut, Lebanon was bombed in 1983, was the only assigned physician and most of the corpsmen. One of the officers who stepped forth to oversee the emergency patient triage, treatment, and evacuation was a Navy entomologist.

Although DoD's primary entomological role is to support military operations the 1950 enactment of the Catastrophe Aid Bill meant resources could be used to assist in the control of civilian epidemics. Navy Vector Control Teams provided assistance after the Great Kansas City Flood (1951); fly control during polio epidemics in South Florida; and mosquito control during St. Louis Encephalitis outbreaks. A combined service effort was used during the 1971 Venezuelan Equine Encephalitis outbreaks, when U.S. Army entomologists provided surveillance of mosquito populations and the U.S. Air Force Spray flight conducted aerial spray missions. Outbreaks of economic pests, primarily locusts and grasshoppers, have been aerially sprayed both stateside and overseas. Many other missions, too numerous to elaborate here, have been conducted to suppress both disease vectors and economic pests.

All of which brings us back to the opening scenario in this article and to the Armed Forces Pest Management Board. In recognition of the developing international role of the armed forces, the Board was established in 1957 to develop and coordinate an interagency approach to controlling insects and related pests of medical and economic importance. A dozen Army, Navy, and Air Force Commands were appointed to the original Board with 9 liaison members from U.S.A., Canadian, and British agencies. Since then, there has been a name change, redesignation of titles and duties of key personnel, reassignment within DoD, and expansion of the mission. Today's Board develops and recommends DoD policy, provides technical and scientific advice to DoD Components, coordinates DoD pest management activities, works with a Pesticide Hot Line (operated by the US Army Environmental Hygiene Agency), sponsors the Defense Pest Management Information Analysis Center (DPMIAC), and provides liaison officers to coordinate entomology research and contingency efforts worldwide. This unique organization is composed of medical and engineering pest management professionals from each of the military departments and over 40 other military agency representatives. Approximately 35 other federal organizations and allied defense agencies are represented by liaison members.

Information is the lifeblood of today's computerized DoD entomologist. To this end the DPMIAC maintains a library and automated bibliographic database, and also accesses the major technical information resources. The Center provides each DoD entomologist with a comprehensive bimonthly Technical Information Bulletin. The DPMIAC has developed a Global Pesticide Resistance Inventory and a series of Disease Vector Profiles which describe the arthropod-borne diseases and vectors present in most countries of the world.

The military has vigorously pursued entomological research and development in the areas of vector biology/ecology/control, epidemiology of arthropod-borne disease, resistance, methods of protecting individuals and groups from arthropod-borne disease, improved insecticides/repellents and improved equipment/ techniques for the dispersal of these chemicals. The Army has research operations at the Biomedical Research and Development Laboratory and the Medical Research Institute for Infectious Diseases, both located at Fort Detrick, MD and Walter Reed Army Institute of Research, Washington, DC in the US, as well as overseas sites, including Thailand, Malaysia, Brazil, and Kenya. One of the most recent and notable Army R & D efforts has been the development of a longer lasting DEET repellent lotion, and a new repellent treatment for uniforms. The Navy stations research entomologists in locations abroad including Peru, Egypt and Indonesia. But the needs are greater than in-house resources can fulfill, so the excellent relationship begun with the USDA Orlando Laboratory during WWII has been perpetuated and expanded. Today several USDA labs provide support for DoD research needs regarding disease vectors, stored products pests and wood-attacking insects. Universities and civilian contractors satisfy still other research needs. Over the years, DoD entomological research has seen considerable success in developing materials and strategies to protect operational forces from endemic vector-borne diseases.

A Look Ahead

Looking into the future, we will find that the mission of military entomology will remain: Conserve the Fighting Strength. To this end, our efforts will focus on the need to protect susceptible troops, who must deploy anywhere in the world on an instant's notice, from arthropod-borne disease. However, we can not lose sight of the fact that we must also maintain the facilities and other resources we use to train and outfit these troops. One of the greatest challenges to our entomologists will be to accomplish this task in a period of heightened environmental concern, stricter tightening pesticide regulations and tightening budgets. To meet this challenge innovative IPM strategies, some being developed today and others perhaps not yet imagined, will be needed.

Malaria and Mosquitoes in a Mexican Town

Earle S. Raun

"Welcome!! Welcome!", in Mexican, were the words I heard as I first set foot in the small village of Tamisco. As I alighted from the USDA Chevrolet panel truck, the chief of the village greeted me. Welcome was not the case when entomologists arrived there 2 years previously to supervise the beginnings of a project that led to an eventual World Health Organization's malaria control program. Instead, suspicion was rampant among the villagers. Mexican police accompanied the workers of the entomological team.

In 1947, as a new research entomologist with the USDA, Man and Animal Insects Laboratory at Orlando, Florida, I was fortunate to be designated to help in a Malaria control research project. This project was a joint effort of the USDA, Rockefeller Foundation and the Mexican government. Carried out over a 6 year period, my involvement was during January, February, March, and April of the third year.

Malaria is a disease caused by a protozoan in the blood. It is transmitted by mosquitoes of the genus *Anopheles*. A world-wide problem of tropics and sub-tropics, it had been a scourge of American forces in the pacific during World War II.

The adult malaria-carrying mosquito rests indoors on the walls of dwellings during the day, and takes a blood meal from humans sleeping in the dwelling at night. As it feeds, it injects salivary fluid into its host. The salivary fluid carries the malaria microbe and infects the host.

Scientists had discovered DDT as an insecticide in the 1930s. During the War years, its residue was found to kill mosquitoes for about a year, if they rested on a treated surface. Laboratory tests indicated that malaria could be eliminated as a human problem if the resting surfaces in towns and villages were to be treated with DDT. In the winter of 1945 entomologists from the USDA, medical teams of the Rockefeller Foundation, and workers provided by the Mexican government began the 6 year project.

Six small villages in the mountains south of Cuerna Vaca, Morelos, Mexico, were selected for the test, with 3 of them to be treated and 3 to act as controls. The medical teams took medical records from the population, measuring rates of various diseases, including malaria. Entomologists observed the mosquito populations, as well as the incidence of bed bugs, lice and flies. Then the Mexican workers, supervised by an entomologist, applied DDT to the insides of all of the buildings in the 3 designated villages once each year for three years.

My involvement the third year was to supervise the DDT applications. It led to a very interesting and satisfying period of my career, for many reasons.

One of the reasons was the obvious pleasure shown by the residents of the three villages we were treating. They informed me that the previous treatments had so changed the comfort and health of their lives that they wanted it to continue. Their homes were no longer full of bed bugs, cockroaches, fleas, spiders, and scorpions.

Another reason was that the medical teams were documenting a dramatic reduction in malaria and dysentery. In other words, the letters, DDT, meant comfort and better health to these simple, hard-working, poor people.

Interest was provided entomologically and sociologically. The experience gave me a chance to see many of these medically important insect populations in their natural state, rather than in a laboratory. Their voracious feeding, ability to reproduce, and the diseases they carried were much more impressive than any artificial situation could be.

Sociologically, this mid-western reared and trained young entomologist had never been exposed to the poverty rampant in much of the world. Observing their helpful attitudes to one another, their pleasure in simple things, and their struggles to survive, made me wonder at the justice (or injustice) of their lot.

The success of the project, at its termination, was evidenced by the adoption of this approach to malaria control world-wide by the World Health Organization. The subsequent discovery that DDT had disastrous environmental consequences to many forms of animal and bird life was a great disappointment to many people around the world, whether they knew it or not. Malaria and *Anopheles* continue.

Mosquito Control – Its Impact on the Growth and Development of Florida

John Andrew Mulrennan Jr.

In 1845, when the Statehood of Florida was being debated in the Congress, the Honorable John Randolph of Virginia rose to declare that Florida could never be developed, nor would it ever be a fit place to live. He described it as "A land of swamps, of quagmires of frogs and alligators and mosquitoes." This was a perfect description at the time when the population was 66,500, the greatest percentage of which was living the northern part of the state from Pensacola to Jacksonville.

Human existence for many during those days was bare survival due to the ravage of parasites and diseases that plagued the people on a year-round basis. The most horrifying of the diseases was yellow fever which swept through the towns and villages in the summer months affecting literally hundreds of people.

Great pestilence

The year 1857 is remembered as a time of great pestilence. It was reported that approximately 600 persons in Jacksonville had yellow fever and that 127 died.

In 1877, 1,612 out of a population of 3,000 in Fernandina were sick with fevers. Ninety-five people died including Dr. Francis Preston Wellford, president of the Florida Medical Association. Dr. Wellford had volunteered to aid the sick and dying and left Jacksonville for Fernandina on September 22. Seventeen days later, he died of yellow fever.

In 1887, yellow fever epidemics occurred in Key West, Tampa, Plant City and Manatee. In 1888 the disease broke out in Jacksonville causing an exodus from the city of people in carriages, drays and wagons streaming toward the depot and docks while every outgoing train and steamer was crowded beyond capacity. In all nearly 5,000 had contracted the disease and more than 400 had died. The epidemic finally subsided on November 25 when the temperature dropped to 32 degrees.

For a period of 141 years epidemics can be chronicled throughout the state with six major pandemics culminating in the great pandemic of 1888. During the 141 years well over 25,000 cases were recorded with more than 5,000 deaths.

It was the pandemic of 1888 that provided the final impetus for establishing the State Board of Health and naming of Dr. Joseph Y. Porter as the first state health officer. Earlier attempts had always failed, e.g., in 1873 a bill which carried a $200 appropriation was introduced in the legislature to organize a State Board of Health but was defeated on the basis of being too expensive.

Dr. John P. Wall in an 1875 address to the Florida Medical Association stated: "The time is fast hastening when preservation of the public health will become one of primary consideration in all enlightened governments." This statement laid the groundwork for the advancement of public health. Dr. Wall is often called the father of the State Board of Health as a result of his early efforts. He was a man far ahead of his time because in 1873 he postulated that the common "treetop" mosquito was the carrier of yellow fever 27 years before Carlos Finley proved it.

Malaria belt

While the occurrence of yellow fever epidemics were visibly dramatic and devastating, the occurrence of malaria and dengue was far more subtle but, over the years, equally as devastating.

The early settlements of Florida were almost entirely within the area later defined as the "malaria belt." Tallahassee, the state capitol, was in the midst of this region. Malaria persisted well into the 20th century affecting thousands of Floridians. This disease, along with dengue and yellow fever, caused retardation of Florida's growth and unestimatable economic damage.

In 1898 Sir Ronald Ross proved the role of the *Anopheles* mosquito in transmitting malaria and in 1900 Walter Reed verified the mosquito theories of Wall and Finlay relating to yellow fever. These discoveries marked the beginning of the end of the three major mosquito-borne diseases in Florida. The last case of yellow fever occurred in 1910, dengue in 1932, and malaria in 1948.

The discovery of the mosquito's role in disease transmission lead to the beginning of mosquito control programs in Florida. Major emphasis was placed on elimination of breeding sources. The elimination of cisterns for drinking water, rain barrels, and other receptacles eliminated the vector and yellow fever.

Control efforts

Malaria control efforts were first organized during World War I when the U.S. Army, U.S. Public Health Service, and State Board of Health set up a program of drainage and larviciding at Camp Johnson near Jacksonville. In 1919 the state, city of Perry, and

Burton-Swartz Cypress Company jointly set up a malaria control project in Perry, one of the state's most highly malarious areas.

All the early mosquito control efforts in Florida were aimed at disease control but as early as 1922, when the Florida Anti-Mosquito Association was formed, the importance of pestiferous mosquito control was becoming apparent. Enabling legislation for creation of mosquito control districts was passed in 1925 and in the next ten years five districts were formed. Their main efforts were directed toward the salt marsh mosquito.

In the ensuing years, local, state, and federal agencies, and private foundations such as the Rockefeller Foundation combined research and control efforts to bring about elimination of malaria and dengue and reduction on a grand scale of the pestiferous salt marsh mosquitoes.

Most people will agree that it has been the control of mosquitoes and mosquito-borne diseases that has allowed Florida to expand its agriculture, increase tourism from a winter business to year-round and provide a mecca for millions of people to work and play. Florida is now one of the fastest growing states in the nation.

New problems

While the economic benefits of Florida's growth is enormous, it presents new problems in public health and mosquito borne-disease control. Florida has changed since the turn of the century from an essentially rural state with small villages and towns to large urban and suburban areas. These areas are often situated in or near large mosquito breeding areas or, by their construction, have created new ones.

St. Louis encephalitis (SLE), an urban/suburban vector-borne disease, now poses a threat to our population Epidemics occurred in the Tampa Bay area in 1959 (68 cases, five deaths); 1961 (25 cases, seven deaths); and 1962 (222 cases, 43 deaths). In 1977 another epidemic occurred involving 23 counties in central and south Florida with 110 cases and eight deaths. Isolated human SLE cases have occurred occasionally since then and in recent years the virus has been routinely isolated from sentinel chicken flocks primarily in south Florida.

Eastern equine encephalitis (EEE) also occurs sporadically in humans and equines and human cases are usually associated with campers, hunters or fishermen – who have frequented fresh water swamps that have a history of endemic EEE.

It seems now that there is an irony associated with our mosquito and vector-borne disease control programs. We have successfully eliminated yellow fever, malaria, and dengue and greatly reduced pestiferous mosquitoes. This has allowed Florida's population to grow. The growth that has occurred has created new man-made mosquito habitats conducive to vector mosquito breeding or has placed human populations in close proximity to endemic EEE areas. These situations pose potentially serious public health problems in Florida's future.

History often repeats itself. We can only hope that this is not the case for mosquito-borne diseases in Florida. Florida's history of these diseases is too terribly devastating to be repeated.

Used with permission from J. Florida Medical Assoc. 73(4):310-311 (1986)

In Pursuit
of a Better Repellent

Carl E. Schreck

The USDA Insects Affecting Man and Animals Research Laboratory (IAMARL) of Gainesville, FL, had its origins within four months after the United States entered World War II. Until 1963, it was located in Orlando, FL, and was popularly called the Orlando Laboratory. Its purpose was to develop the ways and means to protect our combat troops from disease-bearing arthropod pests. In relatively austere surroundings, with a very limited data base available for those times, a group of pioneering entomologists achieved some conspicuous milestones in the control of medically important arthropods.

One of the high priority and quickly organized projects was concerned with personal protection. In charge of the project were B. V. Travis and F. A. Morton. "The need for a good repellent is considered urgent," said E. F. Knipling, Laboratory Director during the critical war years. To this day, the concepts and procedures developed by these and other distinguished scientists continue to be sound approaches in the identification and investigation of arthropod repellents.

From 1942 to the present time, tens of thousands of chemicals have been assessed for repellent activity. With all due respect to those entomologists and chemists who have diligently worked on this problem, the ultimate repellent has yet to be discovered. The most effective and widely used all-purpose repellent is deet. Deet was identified and developed at the IAMARL in the mid 1950s when the laboratory was under the direction of C. N. Smith. Unfortunately, deet has several widely recognized undesirable characteristics. These include oiliness, odor, and a propensity to dissolve some paints, plastics, and synthetic fabrics. And some pests, such as deer flies, are relatively unaffected – so the search continues.

In the course of the search for better repellents, humorous things have happened, and I would like to share some of them. Though these events have occurred since I have been associated with IAMARL, I am certain Dr. Smith (a far better story teller than myself) and others who have retired, could provide further accounts of incidents that turned many a wearisome, humdrum day into an amusing memory. In any event, here goes.

Since the 1950s we have tested large numbers of experimental repellents including various chemicals, mixtures, plant extracts, and essential oils. Needless to say, some had very objectionable odors. Most washed off easily, while others persisted even after numerous scrub-downs. Upon arriving home after working with these materials, the story was always the same. Our wives would meet us at the door, say "phew," escort us to the shower and throw our clothes in the wash.

On one occasion some years ago, one of the men, still carrying traces of a noxious odor, was in the checkout line of a food store. He was standing in front of a bearded young man with a ponytail and clad in tattered jeans typical of the style of the day. A lady in front turned on several occasions and looked disapprovingly at the young man. Finally, as she was leaving, she turned, wrinkled up her nose and said, "You would think those hippies would know enough to bathe!"

Another time, same scenario, a technician exuding a persistent and embarrassing chemical odor was leaving with his groceries and as he walked away the bag boy said to the cashier, "P...U, someone just bought some rotten groceries."

Once, after having worked with some particularly malodorous materials, one of the men had to leave the lab early to bathe, change clothes, and take his mother to a funeral. Later, when they were outside, she whispered, "Son, that was a very nice service and all, but somebody in that church sure did need to take a bath."

Then there was the time we were testing tick repellents south of Savannah, Georgia, in a huge forested area. Our project leader, who had a reputation for being gruff, warned us as usual about poisonous snakes and getting lost (he prided himself in knowing his way around this forest he had visited for so many years). We went into the woods to find "hot spots" where ticks are concentrated in large numbers. We would sit in these spots, until exposed sufficiently, to learn if the repellent-treated clothing we wore was providing satisfactory protection. In looking for tick infestations, we walked some distance, and it was common to get separated.

On this occasion, after the usual exposure period, I began to head back toward the rendezvous site when I heard hooo.., hooo.., hooo.., from the project leader deep in the forest. Believing this was a signal that the test period was over, I called back "okaaayyy.." and received another three hooo's, seeming to confirm it. When I finally arrived at the meeting place, everyone was there except the project leader. Because it was getting late, we began to wonder where he was and honked the car horn several times. Further time passed until finally we heard someone crashing through the brush coming our way. Suddenly, there he was, puffing hard, red faced and obviously furious. He immediately began chewing out the nearest technician for not answering his call. With a sheepish grin, the technician said he thought it was a hound bellowing. I thought to myself, he had *not* been signaling the test was over, he had been calling because he was lost! I later realized his hooo, hooo, hooo, was really help, help, help, in disguise. He had wanted someone to answer so he could track the voice out. We never told him we knew.

On another occasion, six entomologists and technicians were in New England testing clothing treated with various chemicals against the major tick vector of Lyme disease, *Ixodes dammini*. As a precaution against the disease we gave everyone the option of wearing blue pantyhose under their field clothing to help prevent tick attachment. At first the men did not want to wear the pantyhose but later changed their minds when the hose were renamed "Titanic-Tick-Tights." We were a funny looking lot when we removed our outer clothing after a test exposure to examine it for ticks. Needless to say, some very uncomplimentary remarks were passed around.

The worst experience during this field study was being the untreated control. Not only was one exposed to tick bites, but also took the chance of contracting the disease. An untreated control is essential in this work and so each day, without exception, a different person was assigned the job. No one relished being the control. When it was the entomologist's turn, it was a high point for the technicians – after all, we had designed this insane test.

The day came when one entomologist, not very experienced with the procedure, was scheduled to be the control. He appeared a bit uneasy about having large numbers of ticks crawling all over him. That day it was difficult for us to find "hot spots" to sit in to get infested so it seemed to put him at ease. After we had visited several areas and had not found many ticks, one might say he was almost enjoying himself. Suddenly, from behind some shrubs he shouted, "My God, get them off me, get them off me," and began leaping about in a frenzy, slapping at his clothing. He had exposed himself to larval ticks, commonly called seed ticks. All at once he had discovered that hundreds of the tiny blood suckers were crawling all over him. They are small, so sometimes they may go unobserved by even a practiced eye, and the bite can be just as distressing as that from an adult. Pretending to be upset that he was deliberately reducing what could be the highest control count we had had all week, I admonished him for knocking the ticks off his clothing and for setting a poor example as a professional entomologist in the presence of the technicians, who were sporting wide grins during all this. Well, as it turned out, he had a good control count of ticks after all.

When we go to the field, it's often to remote areas and we have to brown-bag it. Most people have fond memories of picnic lunches in the great outdoors, but not us. Our picnic sites are ones you would choose for your worst enemy. The reason of course is that field testing must be done in places where target insects are at high density. After all, who would want to use a repellent that is *only* effective when there are just a few mosquitoes around? Anyway, at noon we do the best we can. If we all get into the car to eat our lunches, there is not enough room, or it's hot because we must keep the windows closed to prevent the mosquitoes, black flies, biting midges, or deer flies from entering, too. The only way to eat lunch in some comfort is to wear protective clothing. Did you ever try to drink an RC Cola through a head net?

One hot, humid day we were sitting outside just finishing a rather dull lunch when one fellow unwrapped a big wedge of chocolate cake. He took a bite, savored it, and, I am sure, pretended he was somewhere else, while hungry mosquitoes flew all around him. As he sat there, poised for another bite, a figure moved toward him out of the brush. He didn't spot it until it was fairly close and it gave him a start. Coming at him was the ugliest three-legged raccoon you ever saw. "Shoo, go-away, go on, git!" The raccoon kept coming, oblivious to the shouted threats, its eyes focused on the chocolate cake, with a cloud of mosquitoes as an entourage. " Shoo," the man said louder, in an uneasy voice. It was pretty obvious to us that this ugly three-legged 'coon would not be intimidated. This was an experienced moocher. It came very close and reached for the cake. Totally unnerved, the man threw it over the animal's head. Swiftly, the raccoon gathered it up and disappeared into the woods along with its companion cloud of mosquitoes. So much for our style of getting back to nature.

Then there was the time in the Everglades National Park when, as usual, we left our lunches in the van while we worked some distance away. On our return at noon, the Park's educated crows were stationed around the van making soft suspiciously pleasant caw..caw..caw.. sounds. Found, instead of our lunches, were empty wrappers scattered everywhere. Our feathered friends had entered the van through a partially open window and eaten all our food! It appeared they had stayed around to enjoy our misfortune, because I could have sworn those crows were smiling as we sullenly piled into the van to go and search for something to eat. Since no one in our party admitted to leaving it open, the assumption was that one of the crows opened the door on the passenger side and cranked the window down.

I suspect the reader is wondering if it is dedication or dim-wittedness that attracted us to this occupation. After all, there are not too many professions in which the duties include *deliberately* exposing oneself to hoards of blood-sucking insects, ticks, and mites. On the other hand, working with insect repellents and other forms of personal protection provides a unique and challenging career. It has bestowed us with a rare opportunity to view many diverse peoples, cultures, and habitats – from near the Arctic Circle to below the Equator and from the Hindu Kush to the origins of the Blue Nile. Perhaps a few mosquito bites isn't such a big price to pay, after all!

Suggested Further Reading

Knipling, E. F. 1948. Insect control investigations of the Orlando, Fla., laboratory during World War II, pp. 331-348. *In* Annual Report of the Board of Regents of the Smithsonian Institution, June 30, 1948 (Publication 3954). Superintendent of Documents, U. S. Government Printing Office, Washington, DC.

Resolving the Screwworm Problem

E. F. Knipling

One of the most destructive and certainly the most obnoxious insect pest in North America was the screwworm *Cochliomyia hominivorax* (coquerel). The adult has a robust body about four times the size of the house fly. It has a metallic blue color with three darker stripes on the thorax. In 1933, E. C. Cushing, U.S. Department of Agriculture, recognized that the species had for years been confused with an abundant scavenger fly very similar in appearance. The true screwworm, however, is a facultative parasite of mammals, including humans. In relative terms, the true screwworm fly exists in very low numbers. This taxonomic discovery led to a new era in screwworm research and development.

The fly deposits a compact mass of about 300 eggs on the edges of any type of wound on its hosts. The tiny larvae hatch in less than 24 hours and begin feeding on the flesh. They grow rapidly and become fully grown in about 5 days when they are then about 3/5 of an inch long and about the diameter of a kitchen match. The grown larvae leave the wound and pupate in the soil. Adults emerge about 8 days later, and mate on about the 3rd day. Oviposition begins on about the 6th day. In warm weather the complete life cycle requires about 3 weeks.

The larvae are closely packed in the wound and rasp the flesh of the host with their strong mouth hooks. Several rows of small dark recurved spines border the segments of the pinkish larvae. Under magnification the larva has the general appearance of a screw, hence its name. Because of the recurved spines and their aggregating behavior in the wounds, the infested animals cannot readily dislodge the larvae by licking or biting.

As infested wounds increase in size more flesh is consumed and an increasing amount of blood and lymph is lost. The strong disagreeable odor is increasingly attractive to other screwworm flies and a variety of scavenger species. If not treated, the wounds within two weeks may have several thousand feeding larvae of all sizes. Animals are literally eaten alive. In areas of high fly densities, infested animals not found are virtually doomed for a slow traumatic death. Those working on methods of screwworm control become highly motivated to find a satisfactory solution to the problem if for no other reason than compassion for the suffering of the animals that become infested.

Livestock growers in the Southwest for years had lived with this serious economic pest problem. They made every effort to reduce losses through management practices that would minimize the number of surgical and accidental wounds on livestock during the fly season. They practiced controlled breeding so that animals would give birth when flies were not present or scarce. The navels of newborn animals are highly susceptible to screwworm attack. Livestock growers also had a variety of smears to repel flies from wounds and to kill larvae in wounds already infested. But these control measures still permitted hundreds of thousands of screwworm infestations in livestock each year. Wild animals were at the complete mercy of the pest. These animals, especially deer, served as reservoirs to maintain viable screwworm populations even if livestock growers minimized screwworm cases by following good screwworm management practices. Animals that have infested wounds tend to seek brushy areas to escape fly annoyance. Cowboys rode ranges looking for infested animals, but many such animals could not be found and treated in time to prevent serious injury or death, thus, many larvae matured that later emerged as adults. I estimate that during favorable conditions fly populations grew by 3-5 fold each cycle. Thus, an overwintering population of as few as one million flies could grow to a billion flies or more by season's end.

The screwworm is, however, highly susceptible to cold weather. Its overwintering areas in the Southwestern States were usually restricted to the southern portions of Texas, New Mexico, Arizona, and California. It survived all year in Mexico and Central America. As weather became warmer each spring the populations began to increase and spread northward. By flight and movement of infested animals the fly could spread up to 1,000 miles or more before cold weather again pushed the pest back to its overwintering areas. The pest became established in the Southeast U.S. about 1933, where it could overwinter in Florida. Its pattern of spread in the Southeast each spring and summer was similar to that in the Southwest.

While engaged in research on the screwworm early in my professional career, I soon realized that the control measures practiced by livestock growers at that time would never solve this serious pest problem in a satisfactory manner. My thoughts were concentrated on preventive measures that might be applied against the adult population in all of its natural habitats before the flies could attack the domestic and wild hosts. Thus, emphasis would be to manage the pest population rather than attempt to manage the hosts. The use of insecticides to kill the adults in several hundred thousand square miles of fly habitats at that time, and even today, would be out of the question. Therefore, other approaches were considered.

One of my colleagues, R. C. Bushland, and I discussed various concepts of screwworm control from time to time in 1937 and 1938, while we were engaged in research on the problem at Menard, Texas. Bushland and his associate, Roy Melvin, had developed *in vitro* methods of rearing the insect. The potential existed for rearing the fly in large numbers at relatively low cost. Also, we knew by the ecological behavior of the fly that the natural adult populations existed at very low numbers and were greatly restricted in distribution during the winter months. Therefore, it was rationalized that it might be practical to rear and release more screwworm flies than existed in the natural population. Then, if some means could be developed for the flies to transmit lethal factors to the native flies though mating, it might be possible to achieve "autocidal" control. The possibility of developing a genetic strain that possessed detrimental traits that would not be critical to fly rearing in captivity but would inhibit survival in the natural environment was one approach discussed. A "spineless" larval strain, like the spineless cactus developed by Burbank, would be an example of such genetic deficiency. This should not hamper rearing on artificial media but it would no doubt be a serious handicap to survival in nature. We also discussed the possibility of achieving sterility of the male flies by chemical means before releasing them to compete with normal males for mating with normal females in the natural populations.

It was not possible, however, to undertake research on the various ideas considered. We were each assigned to other research projects by 1939, and later were engaged in high priority investigations on arthropod pests of military importance during World War II. Nevertheless, I continued to explore the possibility of screwworm control by genetic means. A hypothetical suppression model was developed which indicated that if a given ratio of sterile to fertile flies existed in natural populations and this caused a decline in the wild population, the same number of sterile flies would lead to progressively higher ratios of released to native flies each generation. In theory, this would in time result in complete inhibition of reproduction. It would represent a new mechanism of pest population suppression. All other methods of control known at that time had a constant effect or tended to become less effective as the pest population declines. I analyzed records of screwworm infestations in livestock in Florida, especially during the winter months and concluded that natural fly populations during that period probably did not exceed 20 flies per sq mile per week.

For more than a decade the possibility of achieving screwworm control by genetic manipulation was discussed with a number of entomologists and geneticists. The response, in general, was negative or highly skeptical. But in 1950, A. W. Lindquist, another colleague who knew of the interest in ways of sterilizing male screwworm flies, called my attention to a publication by H. J. Muller on radiation damage to genetic material in *Drosophila*. I wrote to this eminent scientist describing the theory of screwworm suppression by releasing sterile flies in natural fly habitats, and asked for his opinion on the possibility of causing sterility in screwworm flies by radiation. In his response he expressed the view that the screwworm fly would be influenced by radiation damage in a manner similar to *Drosophila*.

Dr. Mueller's encouraging response was discussed immediately with Dr. Bushland from my Washington, D.C. office. He was leader of the U.S.D.A. livestock insect research project underway at Kerrville, Texas. Although no funds were available for such research, the decision was made to divert funds from other projects and initiate research on sterilization of the screwworm fly.

Dr. Bushland obtained excellent cooperation from an army medical unit near San Antonio that had suitable x-ray equipment. With the assistance of D. F. Hopkins various stages of the screwworm were exposed to x-radiation. Within 6 months they demonstrated that the exposure of pupae near adult emergence to 2500 to 5000 roentgens of x-rays resulted in complete sterility of both male and female flies, without apparent adverse effects on the mating behavior of the male flies. When sterile males in cages were placed in competition with normal males for mating with normal females, the ratio of sterile to fertile eggs masses deposited by the normal females was essentially the same as the ratio of irradiated to normal males in the population. Female screwworm flies seldom mate more than one time. These very encouraging results were followed by field release experiments to determine if the irradiated males would perform in a normal manner in natural populations. The first valid test of this technique was undertaken on Sanibel Island, Florida, under the direction of A. H. Baumhover. This 20 sq mile island is isolated from the mainland only by about two miles. Wound reared males that had been exposed to radiation in the pupal stage were released at the rate of 100 per sq mile per week.

Sterile egg masses deposited on wounded goats by wild flies began to appear within a week. By 3 months the number of egg masses collected began to decline and the rate of egg mass sterility stabilized at about 90 percent. It was obvious however, that eradication of the screwworm population could not be demonstrated because of the migration of fertile mated flies from the mainland. However, the results of this experiment clearly demonstrated that the sterility technique offered promise as a new means of insect suppression.

Funds were not available, however, to undertake a field test in the U.S.A. on the scale that would be necessary to demonstrate eradication. Fortunately, however, a routine request for information on screwworm control was received from B. A. Bitter, veterinarian, on the island of Curaçao, Netherlands Antilles. Curaçao consists of about 170 square miles and is isolated from the South American coast by about 50 miles. This gave us an opportunity to undertake a valid eradication experiment against an isolated population with the amount of funds we could muster. An agreement was negotiated with the Netherlands government to undertake such an experiment on Curaçao. We stressed to all concerned that this was an experiment and success could not be assured. Sterile fly releases were started in 1954 under the direction of A. H. Baumhover. Screwworm pupae reared in Florida were irradiated and flown to Curaçao. The emerging flies were distributed by aircraft at the rate of 1,000 per sq mile per week. Within 3 weeks (comparable to one screwworm cycle), about 70 percent of the eggs masses collected on goats were sterile. In the next 3 weeks the sterility averaged 84 percent; and by the third three week period very few egg masses were collected and all those that were collected were sterile. In general, the trend of the natural population followed the pattern that was predicted by a theoretical suppression model developed 10 years earlier. It might be noted that the trial program was undertaken without any publicity. If it had failed we were fearful that certain budget "watch-dogs" and the news media could have had great pleasure in condemning the expenditure of public funds on what had been dubbed a "screwy" idea by some critics. Since the technique involved sexual behavior of a pest it would certainly have been ridiculed at that time.

The announcement of the successful eradication experiment created considerable interest among scientists. However, the most interest was shown by some of the leaders of the livestock industry in Florida as Charles Scruggs has stated. C. L. Campbell, state veterinarian for the State of Florida, J. O. Pearce, President of the Florida Cattleman's Association, Lat Turner, and others began pressing for a screwworm eradication program in the Southeast, using the sterility technique. The pest had created havoc among livestock in the Southeast since its establishment about 1933. As is typical of research workers, we were reluctant to recommend an eradication program without the opportunity to undertake more research, particularly on large scale rearing procedures. A jump from the production of 170,000 flies per week for distribution on 170 sq miles to a program that was estimated to require the production of 50 million flies per week for distribution on 50,000 sq miles, would obviously involve many problems and considerable risk of failure.

But the leaders of the industry were persistent in demanding the initiation of a program as soon as possible.

The transfer of basic insect population suppression technology into large scale operational programs is not a simple matter, however. Small scale experiments involving the use of highly mobile biological agents against highly mobile pests in non-isolated areas do not provide the information research workers and program managers would like to have before recommending the initiation of practical programs. It is very difficult to convince administrators and budget officials that special funds are needed for well planned and executed pilot tests. Therefore, the only alternative is to accept certain risks and inefficiencies in the conduct of operational programs or delay the practical utilization of new and potentially useful technology for an indefinite period of time. Unfortunately, this is as true today as it was 40 years ago.

An incident occurred during the early stages of the discussions on the possibility of eliminating the screwworm from the Southeast which made a lasting impression on my thinking on this matter. A. W. Lindquist, who was Chief of our research program on Insects Affecting Man and Animals, and I were reviewing the research underway at Orlando, Florida. C. L. Campbell, who knew we were in Florida, came to see us. He urged us to visit with Leroy Collins, Governor of Florida, for the purpose of briefing him on the technical feasibility of eradicating the screwworm from Florida using the sterility technique. He arranged for a private plane to take the three of us to Tallahassee. In our discussion with the Governor we expressed the view that the sterile flies should perform in Florida as well as they did on Curaçao. We suggested, however, that additional research be undertaken on several aspects before considering an eradication program. He asked how long this would take and how much would be saved if the research were successful. We proposed a two year research program and estimated that the ultimate cost of a program might be reduced by about $2 million. His comment was that a two year delay did not seem justified with the expectation of saving only $2 million, when the losses due to the pest at that time were estimated to exceed $10 million each year. We did not have a good response to this pragmatic analysis.

We did however, have the opportunity to undertake some much needed additional research. A rather makeshift rearing facility was used that had the capacity to produce about 2 million flies per week. A release experiment was undertaken in an area of 2,000 sq miles in Florida. Since the trial area was not isolated we predicted that maximum sterility in the center of the release area probably would not exceed 50 percent. But within several months sterility reached 70 percent near the center of the release area. Since this exceeded our expectations, the trial was judged sufficiently successful to justify the development of plans for an eradication program. The program was to be financed by the USDA, the State of Florida, and the livestock industry. A screwworm rearing facility having the capacity to produce 50 million flies per week was constructed at Sebring, Florida, by the State of Florida. It was designed by C. N. Husman, USDA, an innovative equipment engineer. Rearing insects in such numbers was the first major advance in mass insect rearing technology.

It would require a volume to fully discuss details of the eradication plans and operations. Publications by Meadows and others suggested for further reading discuss some of the details. The program was started on a small scale in 1957 and became fully operational by mid 1958. The strategy was to release sterile flies from airplanes throughout the expected overwintering area, at the rate of 1,000 per sq mile per week. A cadre of 60 livestock inspectors monitored screwworm infestation among livestock. The screwworm population never approached its normal abundance after the program was initiated. Within about 18 months eradication was achieved.

The managers and operators of the program were faced with, and had to resolve a number of technical and operational problems in conducting the program. Almost every aspect of the program involved technical and operational problems that had not been encountered by previous experiences. The program was directed by R. S. Sharman, USDA veterinarian. Dr. Bushland gave technical direction to the program. The success of the program was lauded by the livestock industry in the Southeastern U.S. It opened a new approach to insect control, and stimulated research on other insect problems that might be amenable to solution by similar means.

The most immediate impact, however, was pressure from livestock interests for a similar program against the screwworm in the Southwestern U.S. The possibility of resolving a pest problem that had plagued livestock growers for many decades lead to virtual demands that such a program be initiated. Drs. Bushland, Lindquist, and I, as well as others, had serious reservations over the technical and operational feasibility of a successful eradication program in the Southwest. Unlike the Southeastern region the fly population in the Southwestern U.S. was not isolated. There were no hopes at that time for a program on the scale that would permit the exposure of a large portion of the natural fly population in adjacent Mexico to sterile flies. It is, of course, axiomatic that adequate suppression pressure must be applied in areas larger than the normal flight range of a pest to expect progress toward eradication. The only hope that we could visualize was the possibility of eliminating the flies from the U.S. and then establish a permanent sterile fly barrier along the Mexican-U.S. border, to prevent the re-establishment of the pest in the U.S. The research staff explained the technical differences in an eradication effort in the Southwest versus the Southeast to appropriate officials within the USDA, the State of Texas, and to representatives of the livestock industry.

Despite these differences, representatives of the livestock industry, including Charles G. Scruggs, T. A. Kincaid, and Dolph Brisco (later governor of Texas), expressed their willingness to accept the risks and uncertainty of such a program. Like the Southeast, the livestock growers in the Southwest took the initiative in pressing for a screwworm suppression program. After months of negotiations on many matters a decision was finally reached to undertake such a program. It eventually involved Texas, New Mexico, Arizona, and California. The book by Scruggs, listed in this publication for further reading, records the magnitude of the problems that had to be resolved before the decision was reached.

The program was under the direction of M. E. Meadows of the Animal and Plant Health Inspection Service (APHIS). The proposed strategy was to make sterile fly releases in the suspected overwintering areas at the rate of 200 to 1,000 per sq mile per week. If eradication proved successful in the U.S.A. a permanent sterile fly barrier 100 miles wide would be maintained along the U.S.A.- Mexico border to prevent the reintroduction of native flies. At the time screwworm biologists thought that the flight range of the flies would unlikely exceed 50 miles. A fly rearing plant was constructed in Mission, Texas that had the capacity to produce 100 million flies per week. This plant was also designed by C. N. Husman. The program got underway in Texas in 1962. During the winter of 1962-63 the results seemed excellent. Very few infested animals were recorded in Texas by late winter. However, by early spring relatively few but widely scattered screwworm infested animals were found in the fly release area and up to 100 miles or more north of the sterile fly release area. This was a great disappointment. Opinions on the reason for the widely scattered screwworm cases varied. The prevailing views were that some flies survived the winter further north than anticipated; or that some infested animals were being shipped across the livestock quarantine line. The possibility that some fertile mated female flies from Mexico had migrated up to 200 miles or more into the U.S. was given little credence by most of those involved in the program. However, our screwworm biologists felt that the possibility of long range flight of prior fertile mated females from Mexico should not be ruled out. It was later shown by fly distribution studies that some marked sterile flies could move at least 180 miles. Also, circumstantial evidence indicated that some wild flies must have the capability of migrating up to 300 miles.

The screwworm infested animals that were found each spring tended to follow the pattern observed after the first winter. Thus, entomologists again underestimated the flight range of an insect species. This meant that the program would have to be recognized as a suppression rather than an eradication program. Nevertheless, the results were outstanding. For the first time livestock growers were experiencing virtual freedom from screwworm infestations in livestock. I estimate that during the first 8 years screwworm cases in the Southwestern U.S. were reduced by about 98 percent.

By the early seventies, however, there were indications that the number of screwworm cases each year was increasing. In 1972 the number of cases increased to an abnormally high level. This caused great concern to livestock growers, program operators, and the research staff. Self proclaimed experts on the screwworm were then being heard from. Some were claiming that the program was a failure, and that it was apparent the pest had developed resistance to the sterility techniques. Others advocated termination of the program on the grounds that it was a waste of money. It should be noted, however, that most of these critics had little or no first hand knowledge of the screwworm problem and had no research experience with the sterility technique.

The research staff undertook investigation on various possible explanations for the decline in the degree of control that was being achieved. My assessment was that the most likely cause for the progressive increase in screwworm infested animals was due to a gradual decrease in the ratio of sterile to fertile flies in the natural populations. During the first few years following initiation of the program the results were so striking that livestock growers saw no need to maintain personnel to look for and treat infested animals. They began to ignore management practices that for years had kept susceptible animals at a minimum during the fly season. The overall livestock population had increased. The deer population probably increased several-fold and had become prevalent in much larger areas. Also, the weather in 1972 was especially conducive to screwworm survival during the winter and spring. Additionally, a severe outbreak of the Gulf coast tick occurred in the coastal regions of Texas during the early seventies.

I estimated that the sum total effect of these changes resulted in natural fly populations by 1972 that were 5 to 10 times higher than during the first few years of the program. Concurrently however, the number of flies that were reared for release remained constant or actually declined somewhat. Also to reduce release costs the fly distribution pattern was changed to wider flight lanes.

I estimated that these changes could have gradually reduced the effective sterile to fertile fly ratio by at least 5 fold by the early seventies. I believe that subsequent developments fully supported by views on the causes of the declining results.

Fortunately the livestock growers largely ignored the controversy that developed among the pest management community. They knew how important the program was to their livestock operations. Also, the managers of the program remained undaunted. They continued to execute the program to the best of their ability within the financial resources available to them. If the program had been discontinued in the early seventies as some critics advocated, I predicted at the time that the changes in livestock management practices that had taken place and the lack of personnel to monitor the livestock, coupled with the increase in wild animal populations would result in catastrophic outbreaks of the screwworm throughout much of the U.S.A., and cause staggering losses because of the increasing costs of labor and the increased value of livestock. Fortunately, the consequences of the recommendations of those who took the negative position on the problem will never be known.

It was becoming increasingly apparent, however, that to achieve the original goal, or even to maintain the high degree of control that had been achieved in the early years, it would be necessary to expand the program and deal with the screwworm population in adjacent Mexico in a more positive manner. Livestock growers in the U.S.A. and Mexico supported plans for such expansion. An agreement was finally reached between the U.S.A. and Mexican governments, and the expanded program was started in 1977. A large screwworm rearing facility was constructed and operated by the government of Mexico near Tuxtla Gutierrez, Chiapis. This rearing plant had the capacity to produce 500 million sterile flies per week. Thus, it would now be possible to release sterile flies in a much larger area and in greater numbers where needed.

The success of this program soon became apparent. Established fly populations were gradually being pushed further from the U.S. Within 5 years the screwworm had been eliminated from all of the U.S. and most of Northern Mexico. Within 10 years the pest was pushed south to the Isthmus of Tehuantepec in Southern Mexico. The ultimate goal is to eliminate the pest from all of Central America and establish a sterile fly barrier at the Isthmus of Panama to prevent re-introduction into North America. It is not possible within the space limitations to give due credit to the many able and dedicated people who contributed to the success of the screwworm eradication and suppression programs. The research scientists who gave support to the program, the managers of the program in the U.S.A. and Mexico and the leaders of the livestock industry all made major contributions to the varied and difficult program. Their efforts prevailed in the face of many technical, operational, financial, and social problems. But the rewards have been very great. Billions of dollars have been saved for the livestock economy in the U.S.A. and Mexico. These benefits will continue to accrue so long as the program is maintained. It is estimated that the benefits, annually, exceed the costs by 10 fold or more. A very serious parasite that plagued domestic and wild animals for centuries has been eliminated from North America.

Aside from the economic benefits to the livestock economy the success of the screwworm program can serve as a model for a similar approach to the resolution of other major insect pest problems in the next century. Scientists in the area of pest management have made outstanding progress on a number of ecologically acceptable insect control techniques that used alone or properly integrated could maintain total populations of many pests at levels below economic significance. To accomplish such goals, however, it will be necessary to direct the technology we have against the pest populations and in an organized and coordinated manner. While the strategy of protecting plants and animals as the need arises has a definite place in insect pest management, so does the total pest population management approach. It is hoped that the outstanding success of the screwworm program will result in a more concerted and unified effort by leaders in the pest management field to analyze specific pest problems objectively, rather than by some preconceived philosophy, and in the light of available technology. There is reason to believe that the total population suppression approach will become increasingly feasible for a number of other major insect pests in the future, and by means that will be largely target pest specific. It was a privilege for me to be closely associated with the various screwworm programs from their inception to their conclusion.

With the author's permission I conclude this discussion by quoting a poem entitled "The Screwworm". The author, a cowboy by profession, describes in a biologically realistic manner, the nature of the screwworm problem and what the successful programs mean to the livestock industry. Only livestock men and biologists who were familiar with the pest, and the problems it created for so many years, can fully appreciate Joel Nelson's unique way of describing the nature of this obnoxious and deadly parasite of livestock and wild animals.

The Screwworm

The open range made cowboys
Who were tops at readin' sign
Wild cattle in rough country
Taught 'em how to use their twine

The trail drives made good cowboys

When the night herd took a run
Those boys would have 'em gathered
By the coming of the sun.

But nothing made good cowboys
In all those days gone by

Like the ugly little larvae
Of the stinkin' screwworm fly

Now the screwworm is disgusting
As its very name implies
It's carnivorously eating
'Ere its victim even dies.

From flies to eggs to larvae
And back to flies again
Their chain of life's unbroken
Its a cycle without end.

They'd work north every summer
And they'd stay till killing frost
The cowboy there to fight them
Rancher there to count his cost.

In rabbits, deer, or livestock
Every wound and every scratch
Was an open invitation
For the screwworm eggs to hatch.

When two bulls would get to fightin'
And their heads was skint up some
You could bet before much time had
passed
Ma Screwworm fly would come

When the cowboy came a ridin' by
There'd shore be worms to dope
And shore enuff he'd get 'er done
With one horse and one grass rope.

A favorite place for flies to lay
Was in a cancered eye
The calf's ear that was full of ticks
Was wormy by and by.

And each of these would soon have
screwworms
Workin' in his head
If they wasn't caught by cowboys
Then they'd purty soon be dead.

Smear Sixty-two, Blackwidow,
And E.Q. Three, Three, Five
Just anything to kill the worms
And keep the stock alive.

Was carried in an ol' boot top
Laced with a leather thong

You could smell a cowboy comin'
And still smell him when he'd gone.

That dope would make his head hurt
And sometimes make him sick
But he knew those worms would suffer
worse
When 'ere they took a lick.

Those crawling bloody messy sores
Would test a cowboy's grit
The ones with weaker stomachs
Better drift up north or quit.

A doctorin' wormy cattle
Was an everyday affair
One hundred eighty straight long days
The screwworms didn't care.

The cowboy ridin' fer the brand
Did what the job demanded
On a fifteen dollar saddle bronc,
Doctorin' cattle single handed.

Some days was fifty miles or more
Two horses – maybe three
Catch 'em, tie 'em, dope the worms,
Untie 'em, set 'em free.

Now usually gov'met programs
Are a minimal success
But the one that stopped the screwworm
Has dang shore passed the test.

Cause it pushed the critter southward
And I hope he's there to stay
Here's to the Mission Fly Lab
And the U.S. D. of A.

But let's drink a toast to screwworms
And the hosses they have made
And to their moms the screwworm flies
And all those eggs they've laid.

For they made some damn good ropers
Of some cowboys long ago.
But we've had enuff, By God,
Let's leave 'em down past Mexico!

**by Joel Nelson, 06 Ranch, Alpine, Texas
– 1987**

Suggested Further Reading

Baumhover, A.H., A.J. Graham, B.A. Bitter, D.E. Hopkins, W.D. New, F.H. Dudley and R.C. Bushland. 1955. Screwworm control through release of sterilized flies. J. Econ. Entomol. 48:462-466

Bushland, R.C. 1985. Eradication program in the Southwestern United States pp 12-15. *In* Symposium on eradication of the screwworm from the United States and Mexico. O. H. Graham, Editor, Misc. Publication No. 62, Entomol. Soc. of America. 68 pp.

Bushland, R.C. and D.E. Hopkins. 1951. Experiments with screwworm flies sterilized by x-rays. J. Econ. Entomol. 44:725-731

Graham, O.H. 1985. Editor and Chairman, Symposium on eradication of the screwworm from the United States and Mexico. Misc. Publication No. 62, Entomol. Soc. of America. 68 pp.

Knipling, E.F. 1955. Possibilities of insect control or eradication through the use of sexually sterile males. J. Econ. Entomol. 48:459-462

Meadows, M.E. 1985. Eradication program in the Southeastern United States pp 8-11. *In* Symposium on eradication of the screwworm from the United States and Mexico. O. H. Graham, Editor. Misc. Publication No. 62, Entomol. Soc. of America. 68 pp.

Scruggs, C.G. 1975. The peaceful atom and the deadly fly. Jenkins Publishing Co., Pemberton Press, Austin, Texas, 311 pp.

Forensic Entomology

Bernard Greenberg

Insects on corpses are mute witnesses to the time of death and provide clues that help solve murders. Forensic entomology is about one hundred years old, yet it is one of the newest branches of entomology. A century ago, J.P. Megnin founded the science when he described eight stages in the decomposition of a human body and some of the insects associated with each stage. A body can become like an airport – insects fly in, lay their eggs, develop and leave in more or less predictable schedules of arrivals and departures.

Human bodies decompose at variable rates, even two bodies placed side by side. After a few days the medical examiner or pathologist has no scientific basis for estimating the postmortem period. But a forensic entomologist who knows the life cycles and behavior of carrion insects can zero in on the time of death, so important in apprehending a criminal or exonerating an innocent person. He or she does this by identifying the species on the body, their stage of development and how long it took to reach that stage. He can tell whether a body has been moved from city to country, buried in soil or submerged in water and then exposed, moved from one region to another, or between a cool and a warm place. Different regions and seasons often have their own distinctive cast of insect indicators.

A 13th century Chinese manual on forensic medicine tells of a murder in a farm community and how it was solved by an ingenious investigator. Suspecting that the murder weapon was a sickle, he assembled all the farmers with their sickles laid on the ground in front of them. It was summer, there were numerous flies and some were drawn especially to one sickle which probably retained traces of blood. The investigator was quick to seize upon this clue and he accused the owner of the sickle. Thus confronted, the man confessed. Today forensic entomologists are increasingly called to testify as expert witnesses in murder trials. A growing data base from laboratory and field studies are solidifying the discipline, and actual cases are capturing the public imagination.

Imagine! The star witness might turn out to be a maggot.

To appreciate how "loathsome" maggots become sleuths let's look at actual cases. But first, why the emphasis on maggots? Because blow flies are typically first to the scene of a murder. Put a piece of meat out on a warm summer day and in minutes flies are buzzing around, soon laying eggs. Bodies are usually discovered in the first couple of weeks when these insects are most busy. In that brief time, maggots can reduce the biomass of a body by sixty percent. Beetles usually establish their beachhead later and are less accurate forensic indicators.

Most carrion flies have little interest in living animals and no trouble "getting wind" of their target and zeroing in. Like all insects and invertebrates, they are cold-blooded. When their eggs hatch the larvae grow at predictable rates depending on the species and temperature. Schedules of development have been worked out for some of the common, forensically important, flies. Armed with this information, the entomologist determines the age of the oldest specimens and counts backwards to the time eggs were laid. Knowing the fly's habits, weather, the body's situation and other relevant facts, he can estimate the time of death. Here is how it works in actual cases.

A 15-year old girl was seen leaving a food store in a Chicago suburb at about 3 p.m. on a July day. She was heading across a field to her home just three blocks away, but she never arrived. At eleven the next morning, her body was found in the field. In the stab wounds were first stage maggots of the common bronzy-green blow fly, *Phaenicia sericata*. Given the hatching time required by the eggs and the size of the maggots, at the prevailing temperatures, the murder probably occurred soon after the girl left the store and not that night.

In a Mafia double murder, two bodies were found in the trunk of a car parked in a housing complex, across the road from a forest preserve. The habits of the three fly species that were recovered reflected the contiguous urban and rural environments. The majority of maggots were *circa* 24 hours younger on one body than on the other, suggesting that the two were probably murdered a day apart. One of the flies does not survive winters in the Chicago area and repopulates from the south each spring, arriving in that area in early summer. In this case, the bodies were found in mid-July and the timing of the infestation with this fly fit its biology. If, however, the bodies had been infested with this species somewhat earlier, let's say in April or May, it would have been necessary to postulate that the murders and oviposition had taken place much farther south. As with many such organized crime murders, the chances are slim of apprehending and successfully prosecuting the killers.

The last homicide case is a paradox. It involved no insects. Why then an entomologist? It was a hot July day in Rock Island, Illinois, with temperatures in the 90's. It was 7:15 p.m. and still daylight when the police arrived at the apartment to find a woman's bloodied body on the bed. She had been dead 12 to 24 hours according to the pathologist. The windows and sliding door to the balcony were open.

Given the weather, the general outdoor environment, and the condition of the body, there should have been numerous flies in the apartment, yet there were none. The victim's boyfriend has been accused and is to stand trial. According to the prosecution, he strangled her the previous night after they had been out together. He returned the next evening, let himself into her apartment with a key, opened the balcony door and window to make it appear the work of an intruder, then called the police. With this scenario the police would have arrived before the flies did. The absence of flies could prove to be more significant than their presence.

Maggots and murder are not all that engage the forensic entomologist. Civil suits enlist his expertise and these cases can be as diverse as the habits of the insect pests. Two illustrations will suffice.

A large consignment of dried mushrooms was shipped from West Germany at the end of August. The shipment passed through Canada, entered the United States on September 19 but was not examined until September 28 when numbers of dead moths were discovered. Who was liable for the costly infestation? West Germany? Canada? Or the U.S.? The moth was identified and its life history was determined from the scientific literature. It turned out to be a widespread pest of dried mushrooms in Europe and requires a minimum of sixty days to complete its life cycle. This was far too long a period of time and too large an infestation to have happened en route either in Canada or the U.S. In this case, liability went back to the German distributor.

Finally, a young couple purchased a home warranted to be termite-free by both seller and an on-site professional inspector. The inspection was performed in June and the couple moved into the house in September. Two days later they discovered evidence of an active termite infestation throughout the house. The seller claimed the infestation must have occurred between June and September. Was the seller correct? Again, the biology of the insect provided the answer. Termite populations build up very slowly. The few eggs laid by a young termite queen take an average of two months to hatch and the nymphs take many more months to develop. During all this time, a few toothpicks would probably be enough food. It usually takes years before a termite colony can do recognizable damage. There was no question that the house was infested years before the sale.

Generally, our association with insects has had a long and miserable history. They bother and bite, spoil our food, attack our possessions and literally, make us sick. But for the forensic entomologist, now some of these very pests are recruits to a worthy cause - to serve justice.

Chapter 3

Insects as Enemies of Our Trees and Forests

THE FAR SIDE   By GARY LARSON

"Pull out, Betty! Pull out! . . . You've hit an artery!"

The Gypsy Moth in Its Centennial Year as a Pest

Ralph E. Webb

The gypsy moth was introduced into the United States in the late 1860's; however, the first significant defoliation caused by the moth occurred in Medford, Massachusetts, in 1889, the year of founding of the American Association of Economic Entomologists, a predecessor of the Entomological Society of America. During the next 100 years both the Society and the gypsy moth increased steadily in stature and influence. Thus this can be considered a centennial year both for the moth as well as the Society. The gypsy moth is a highly complex social, economic, as well as a biological problem. The following is an overview of some of the points one must consider if one is to make "sense" of the gypsy moth.

The gypsy moth as a forest pest

The gypsy moth possesses many physical and behavioral attributes that facilitate its survival and population growth, including high reproduction potential, the ability to feed on over 100 species of trees and shrubs, and behavioral traits that enhance survival. It is currently considered the most important forest pest in the Northeast, to which it is now largely restricted. Since the gypsy moth is expanding its range on a broad front at about 15 miles a year, its importance as a national problem will certainly increase. The gypsy moth has infested about 25% of the 260,000,000 acres of hardwood forests east of the Great Plains that are highly susceptible to it. Between 1980 and 1989, Maryland, northern Virginia, northeastern West Virginia, and central Michigan have become generally infested, so that in its centennial year, the gypsy moth is rapidly becoming a truly national pest. In 1942, E. P. Felt predicted that the gypsy moth would defoliate 25 million acres a year if allowed to spread to its ultimate range. The 12.9 million acres defoliated in 1981 by the gypsy moth over its present range lends credibility to Felt's prediction. This translates into a considerable loss of timber now, and a much greater loss in the future.

The gypsy moth as a pest of shade trees

It is where the forest meets the city that the gypsy moth has been and will continue to be the most severe problem. This is due in part to the greater value of a tree in an urban setting. Using the basic formula of the International Society of Arboraculture and the National Arborist Association, a live oak 14 inches in diameter is worth $271 in an unmanaged wooded area; but that same tree is valued at $2,168 in a residential area. Shade trees make up, on the average, 7% of total residential values. Trees have been shown to play a prominent role in the energy budget of a home. While the placement of the trees with respect to the house is important, research has shown a 20 - 25% maximum potential annual effect of trees on energy use in homes compared with similar homes situated in the open. A second study found that energy savings from shading decreased following defoliation of shade trees by insects. Of course, a dead tree will provide little shade, and becomes a menace costing upward of $1,000 to remove. Trees around homes are in some ways more vulnerable to gypsy moth attack than trees in the general forest. USDA-Forest Service findings indicate that gypsy moth egg mass densities can be ten times higher along the forest edge, due in part to the presence of man-made objects such as litter and old tires that serve as secure resting places (refugia). When gypsy moth populations build up to outbreak levels, there are frequently 5 to 10 million caterpillars per acre that migrate over properties and devour nearly every green leaf. Caterpillars are attracted to vertical silhouettes, including houses, and will swarm over them by the thousands, entering open windows and, on occasion, consuming the house plants. An important noneconomic impact of the gypsy moth in metropolitan areas is due to its allergenic setal hairs. In the 1981 mass outbreak, thousands of cases of pruritic dermatitis were reported, especially from school children. A follow-up study found that at least 10% of the entire community experiencing gypsy moth outbreak was affected. Such conditions make the gypsy moth an object of public concern quite apart from the documentable dollar losses it causes.

Several factors combine to make the gypsy moth a controversial and highly "political" insect. First, many concerns about gypsy moth relate to aesthetics and are difficult to quantify. Moreover, gypsy moth population trends cannot be predicted with accuracy. Will a certain egg mass density result in serious defoliation? Maybe yes, maybe no. Will serious defoliation kill my tree? Sometimes yes, sometimes no. Finally, insecticide spraying for gypsy moths often occurs in the areas where we live. Most of us want the strongest control measures that can be safely applied; however, experts may differ over what is truly safe for man and the "environment."

Gypsy moth population dynamics

The gypsy moth is a forest defoliator native to Europe and Asia. In the mid-Atlantic region of the United States, the gypsy moth hatches in mid-April; weighing but 1 mg, it disperses as an air-borne first-instar larva that climbs out onto the tips of branches and catches the breeze, and is blown like a kite for a considerable distance, its "ballooning" facilitated by long hollow setal hairs and a half-meter strand of silk. Upon settling on a host leaf to feed, the caterpillar typically undergoes four (male) or five (female) larval molts, feeding and growing at a leisurely pace in sparse populations or accelerated in competitively dense populations until about mid-June, attaining a size of one gram (male) or two or more grams (females), and then pupating. Adults emerge after about two weeks. The female, flightless and overweight due to eggs, attracts the male using a sex pheromone. The female mates only once (the male several times unless predated), lays one egg mass of ca. 100 - 1000 eggs (average egg mass size decreases with increasing moth density), and dies. The embryo within the egg develops into a fully developed, tiny caterpillar within a month of being laid in July or August, but does not emerge from the egg until the following spring (April or May), after overwintering in the egg in a state of arrested development called diapause.

Gypsy moth population dynamics are viewed somewhat differently in the United States than in its native Europe. R. W. Campbell (Forest Service, USDA) hypothesized that the gypsy moth in North America exhibits "bimodality" consisting of two stable phases ("outbreak" and "innocuous") and two transient phases ("release" and "decline"). When a general area is in an innocuous phase, most individual populations remain low, with those that increase to dangerous levels declining immediately. In outbreak populations, survival of young larvae, older larvae, and pupae are high compared to what occurs in innocuous populations. Predators, especially small mammals, regulate innocuous populations, while disease, lowered fecundity, and starvation regulate outbreak populations. On the other hand, European workers refer to a gypsy moth outbreak occurrence as a "gradation", which is characterized by three phases: population increase, peak population, and population decline; in between gradations the population is said to be in "latency" (this is the same as the innocuous state of Campbell). Eight major gradations were recorded in Spain between 1920 and 1970, with each lasting 3-6 years. One area of Yugoslavia experienced ten gradations between 1880 and 1956. Many areas of Europe and Asia report similar patterns. Weather is felt to modify the size, duration, and intensity of a gradation. Some gypsy moths are always found in certain locations, and such locations are considered foci. Russian workers have suggested that gypsy moth outbreaks are triggered by large scale weather events acting through the host plants, natural enemies, etc. The fecundity of the female varies greatly and is dependent on the phase of the gradation. During the declining phase, disease can be as high as 97-98%. Individual larval health improves as the population level declines. During the population increase phase, death due to disease decreases greatly and survival rates soar. Parasites become more important, but cannot handle the rising numbers of gypsy moths.

Larval behavior

Larval behavior differs greatly in sparse compared to dense populations. In sparse populations, older larvae spend the day resting in protected locations (refugia), feeding at night, and following trails of silk back to their refugia shortly after daybreak. In dense populations, older larvae feed day and night, and pupate sooner than gypsy moths from low density populations. High density pupae are below average in weight, and females emerging from such pupae produce fewer eggs than those from low density populations. Adults may emerge as much as three weeks earlier in dense populations compared with nearby sparse populations.

Damage caused by the gypsy moth

Campbell analyzed the data accumulated from 1911 to 1931 at the Melrose Highlands Laboratory in Massachusetts (collected from 264 plots in New England) on the effects of gypsy moth defoliation on the forest. He found that levels of defoliation depended not only on the numbers of gypsy moths present, but also on such variables as tree species, and that a spectrum of defoliation levels are possible for a particular initial caterpillar density. Favored trees such as oaks are usually more heavily defoliated. Normally, if less than 50% of its leaves are removed, the tree will maintain the remaining leaves until fall, and little harm will come to the tree; but if more than 75% of the leaves are eaten, the tree will set out a new crop of leaves, thus losing next year's important reserves of nutrients. Such weakened trees are susceptible to dieback and even death. Trees are more vulnerable when already stressed by drought, thus defoliated yard trees should be well watered and properly fertilized. Actual tree mortality is often caused by common "secondary" organisms such as the two-lined chestnut borer or the shoestring fungus that can more readily attack weakened trees.

Although the gypsy moth has many hosts, not all tree species are equally nutritious for or preferred by the insect. P. Barbosa of the University of Maryland found that gypsy moth larvae fed more, developed faster, and produced heavier pupae on preferred hosts such as oak, than on a less favored host such as maple. When the gypsy moth defoliates all the trees in an area for the first time, heavy mortality reaching 50% or more typically occurs to susceptible trees. Large specimen oaks are especially vulnerable. However, it generally takes two or more consecutive years of defoliation to significantly increase tree mortality over natural levels in regions where the gypsy moth has been established for a while.

Natural control of the gypsy moth

In Europe, researchers now recognize that a full array of natural enemies, including parasites, predators, and diseases, is required to keep gypsy moth populations in check. Few important natural enemies of the gypsy moth were present when it first became established in North America. To increase these numbers, the U. S. Department of Agriculture has been importing natural enemies of the gypsy moth for over 80 years. A large number of species of parasitic flies and wasps were reared, released and established under this still ongoing program. Parasites kill numerous gypsy moth eggs, larvae, and pupae, but it has never been shown that parasites alone can regulate gypsy moth populations. Numerous predators of gypsy moths exist including mammals, birds, spiders, and certain insects. Small mammals such as the white-footed mouse are thought to be important in keeping low populations down; however, once a gypsy moth population reaches a certain level, predators can no longer keep it in check. Several important diseases of the gypsy moth exist in Europe. Only one disease, a nuclear polyhedrosis virus (or NPV), is consistently important in North America, although outbreaks of certain bacteria have been reported, and a fungal pathogen caused impressive levels of mortality to gypsy moths during the spring of 1989 in New England. (The fungus, previously known from Japan, requires wet conditions to be maximally effective. The spring of 1989 was exceptionally wet in New England.) The NPV becomes important when gypsy moth populations are high, and is the main cause of gypsy moth population collapse. Naturally occurring NPV is of little importance when gypsy moth populations are low. In Europe, several species of protozoans known as microsporidia have been reported to severely suppress gypsy moth populations, but until recently, microsporidia were not known to occur in North American populations. However, cooperative efforts now under way have introduced a limited number of European gypsy moth-adapted microsporidia into North America.

Gypsy moth suppression programs

The U. S. Forest Service participates with interested states in a cooperative state/federal suppression project. The Forest Service provides cost-share funding and technical assistance in support of such state-wide projects that are generally restricted to aerial application of control materials over fairly large forested areas. The program varies considerably from state to state, and not all states participate. Thus many homeowners find themselves left out of organized programs. The most economical and effective way for such homeowners to manage their gypsy moth problem is to unite with neighbors and have the community sprayed aerially. Helicopters are often preferred to fixed-wing aircraft in urban areas. Carbaryl (Sevin), trichlorfon (Dylox), and acephate (Orthene) have all been used in this context, but, as of this writing, most gypsy moth suppression programs are using either the insect growth regulator diflubenzuron (Dimilin), or the bacterial insecticide *Bacillus thuringiensis* (*B.t.*).

As for ground-based spraying to suppress gypsy moths on individual trees, a homeowner will seldom have adequate equipment to protect shade trees from gypsy moths, and will need to hire an arborist who has such equipment. Carbaryl, methoxychlor, acephate, diflubenzuron, natural and synthetic pyrethroids, phosmet (Imidan), and *B.t.* are all registered for this use.

Individual tree treatments for gypsy moth control include the use of systemic insecticides administered by arborists as implants or injections, and the use of physical barrier bands around tree trunks, usually done by the homeowners themselves. Sticky barrier bands are effective if the bands are put in place just before expected egg hatch, and if the bands are maintained through the season. Many affected homeowners place burlap bands around their trees, physically removing and destroying late instar caterpillars and pupae from under the bands each day.

With the possible exception of area wide aerial spraying, none of the above mentioned control methods will affect community wide gypsy moth population dynamics. Such dynamics are driven by factors of weather, disease, etc., over which the homeowner or community has little control. Caterpillars destroyed by homeowners are often replaced by excess caterpillars migrating from nearby defoliated woodlots. The best that we can expect from the above strategies is the protection of individual trees or groups of trees from being significantly defoliated the year of treatment. Fortunately, gypsy moth infestations eventually decline due to natural causes for a period, so that getting one's trees safely through a particular season's attack may be all that is necessary until the gypsy moth returns in its next population outbreak cycle.

Mention of a proprietary product does not constitute an endorsement or a recommendation for its use by the USDA.

Suggested Further Reading

Anderson, J. F. and W. E. Furniss. 1983. Epidemic of urticaria associated with first instar larvae of the gypsy moth *Lymantria dispar* (Lepidoptera: Lymantriidae)., W. L. 1961. Mass insect control programs: four case histories. Psyche 68: 75-111.

Burns, G. A. 1986. Urban tree appraisal: the formula approach. J. Forestry 84 (1): 18, 49.

Cameron, E. A. 1983. Here comes the gypsy moth. J. Arboriculture 9 (2): 29-34.

Campbell, R. W. 1979. Gypsy moth: forest influence. USDA-Forest Service, Agric. Information Bull. No. 423.

Campbell, R. W. & R. J. Sloan. 1978. Numerical bimodality among North American gypsy moth populations. Environmental Entomology 7: 641-646.

Doane, C. C. & M. L. McManus. 1981. The gypsy moth: research toward integrated pest management. United States Department of Agriculture Technical Bull. 1584.

Dunlap, T. R. 1980. The gypsy moth - A study in science and public policy. J. Forest History 21 (3): 116-126.

Felt, E. P. 1942. The gypsy moth threat in the United States. Eastern Plant Board Circular No. 1. 16pp.

Heisler, G. M. 1986. Energy savings with trees. J. Arboriculture 12 (5): 113-125

Moeller, G. H., *et al.* 1977. Economic analysis of the gypsy moth problem in the Northeast. III. Impacts on homeowners and managers of recreation areas. USDA Forest Service Research Paper NE-360.

Payne, B. R., *et al.* 1973. Economic analysis of the gypsy moth problem in the Northeast. II. Applied to residential property. USDA Forest Service Research Paper NE-285.

Perry, C. C. 1955. Gypsy moth: appraisal program and proposed plan to prevent spread of the moth. United States Department of Agriculture Technical Bull. No. 1124.

Rudie, R. J., Jr., and R. S. Dewers. 1984. Effects of shade on home cooling requirements. J. Arboriculture 10 (12): 320-322.

Webb, R. E., *et al.* 1984. Homeowner applications for gypsy moth. Proceedings of the 1984 National Gypsy Moth Review, Charleston, WV, Nov. 26-29, 1984: 110-116.

Forest Entomology

Robert N. Coulson

Forest entomology includes consideration of both the beneficial and negative effects that insects have on forests. Forests are managed for different purposes: timber production, fish and wildlife, recreation, hydrology, and grazing. Insects are of concern to foresters when their activities negatively impact these forest management goals (Coulson and Witter 1984). Forest entomology deals, in part, with seeking explanations for the causes of insect outbreaks and with ways and means of reducing ecological, economic, and social impact of the pest species. In the following sections we examine (i) types of forest insects, (ii) the ways that insects impact on forest resource values, and (iii) the ways that foresters combat the negative effects of insects.

Types of Forest Insects

Forest insect pests occur in several of the insect orders; but the most important representatives are found in the Coleoptera, Lepidoptera, Hymenoptera, Hemiptera, Homoptera, and Orthoptera. For discussion purposes, forest entomologists often classify the pest species according to their feeding preference. In general, forest insects prefer a particular plant species, a certain age-class of the host, and specific anatomical parts of the host. The most prominent groups are defoliators; sapsucking insects; terminal, shoot, twig, and root insects; seed and cone insects; phloem boring insects; wood boring insects; and gall makers. In addition, arthropod pests that affect recreational use of forests (e.g., mosquitoes, ticks, flies, and others) are also considered. Coulson & Witter (1984) provide a general discussion of each of these feeding groups. Berryman (1988) examines in detail representative pest species associated with the feeding groups.

Impact of Insects on Forest Resource Values

Protection of forests from damage by insect pests is a component of overall forest management practice. In addition to insects, foresters are also concerned with plant diseases and fire. There are three general circumstances where forest insects are of concern: in commercial forests, specialized forest settings (forest nurseries, seed orchards, Christmas tree plantations, wilderness areas, etc.), and urban/suburban forests. Trees in each of these settings have different value and the ways and means for reducing the negative effects are different for each type of "forest" condition.

The impact of insects on forests can be viewed from population, community, ecosystem, and landscape perspectives. Insect activities can affect individual hosts in a *population* of forest trees several ways: directly kill it, impair or slow growth, destroy plant parts, physiologically weaken it and thereby make it susceptible to other tree-killing agents such as plant diseases, directly inoculate the tree with a plant disease, and structurally weaken or cause cosmetic defects. At the *community* level, insect activities can influence the species composition and age-class distribution of forest trees. At the *ecosystem* level, insects mediate and perhaps regulate the rates of primary production, consumption, and decomposition. Furthermore, insects participate in the processes that determine structure, function, and rate of change in forest *landscapes*.

Combating the Negative Effects of Forest Insect Pests

Foresters combat the negative effects of insects using a concept known as integrated pest management (IPM). *Integrated pest management* is defined as "the maintenance of destructive agents, including insects, at tolerable levels by the planned use of a variety of preventive, suppressive, or regulatory tactics and strategies that are ecologically and economically efficient and socially acceptable" (Coulson & Witter 1984). *Suppression* deals with actions taken to regulate or modify populations of pest insects. An example of a suppression tactic is the application of an insecticide. *Prevention* deals with actions taken to minimize risks of adverse impacts of insects on resource management goals. An example of a prevention tactic is harvesting a vulnerable age-class of a forest tree species before the insects consume it. *Regulation* deals with legal statutes designed to restrict or contain the spread of insect pests. An example of a regulatory tactic is quarantine of infested nursery stock. Furthermore, damaged trees may retain value subsequent to infestation by insects and integrated pest management also includes use of such material. *Utilization* deals specifically with actions taken to make use of resources damaged by the activities of the insect. For example, trees killed by bark beetle species (Coleoptera: Scolytidae) can be used in a variety of wood products for a considerable period of time following infestation by the insects.

Forest protection problem-solving and decision making involve integration and interpretation of information on insect population dynamics, forest tree population dynamics, various treatment tactics and strategies, and assessments of impact. Computers are used to assist the forester in making decisions and solving problems. These computer programs, generically known as decision support systems, use the different types of information needed for IPM, i.e., knowledge based on experience of foresters and pest management specialists, technical information derived from research, and historical data. Efficient use of available information is one principal task of effective IPM.

Literature Cited

Berryman, A. A. 1988. *Dynamics of Forest Insect Populations.* Plenum Press. New York.

Coulson, R. N. and J. A. Witter. 1984. *Forest Entomology.* John Wiley & Sons. New York.

Insect Pathology in Canada

J.C. Cunningham

Entomologists in the Canadian government have been enthusiastic supporters of biological control since the turn of the century. Before the Second World War, little attention was paid to insect pathogens and most efforts were focussed on importation and release of parasites. A notable exception was the work of Alan Dustan and F.C. Gilliatt, both entomologists in the Dominion of Canada Department of Agriculture, who published on the use of the fungus *Entomophthora sphaerosperma* for control of the apple sucker and other insects in orchards in the Annapolis Valley of Nova Scotia. Also, Mike Timonin published on the effect of *Beauvaria bassiania* on the Colorado potato beetle in 1939 while working with the Dept. of Agriculture. He later joined the faculty of Carleton University in Ottawa. Although retired, Mike still attends scientific meetings and actively promotes the use of microbial insecticides. The earliest Canadian publication on insect pathogens concerning a bacterium in locusts, dated 1917, is that of E.M. du Porte and J. Vanderleck, working at MacDonald College of McGill University, Montreal.

A European spruce sawfly outbreak in Canada in the 1930's and the subsequent population collapse, presumably due to the accidental introduction of a nuclear polyhedrosis virus along with imported parasites, had a profound impact on the forest protection agencies. The European spruce sawfly was first observed in the Gaspé region of Quebec in 1930 and by 1935 6,000 square miles of eastern North America were infested. It threatened to destroy commercially important spruces. Numerous parasite importations were made from Europe and by 1942 the pest population was in decline. A disease had been observed in laboratory populations in 1936 and in field populations in 1939. Robert Balch and Frederick T. Bird published in 1944 on the pathogen which they believed was a virus. Ted Bird, based in Fredericton, New Brunswick, submitted his doctoral thesis on this pathogen. J.J. de Gryse, Chief of Forest Service Investigations in Canada was greatly impressed by the decline of the European spruce sawfly; populations dropped to an endemic level at which they have remained. Another major insect outbreak occurred in Canada; this time it was the spruce budworm, a perennial problem in North America. J.J. de Gryse felt that he now knew the formula for forest insect control and decided to assemble the very best team of insect pathologists he could muster.

At the same time, Ed Steinhaus at Berkeley was also trying to recruit top scientists and both Steinhaus and de Gryse had their eye on Gernot Bergold at the Kaiser Wilhelm Institute of Biochemistry in Tubingen. Bergold was part of a group that had published on insect viruses in 1942 and 1943; the 1943 paper was a 55 page review on polyhedroses of insects. Ed Steinhaus died before he could write the chapter on Canadian involvement in insect pathology in his book "Disease in a Minor Chord" on which he was working. However, there are some amusing references to Bergold and de Gryse in this book. Steinhaus' efforts to hire Bergold were thwarted by U.S. government red tape. The Canadians, being more adroit in being able to hire aliens — particularly biologists — succeeded. The plot appears positively comical when Steinhaus notes — It was sometime in 1946 that de Gryse, disguised in an army uniform, as was necessary for North American visitors in those immediate postwar days, visited the bomb-damaged Institut fur Angewandte Zoologie in Munich.

Research on the pathology of forest insect pests owes much to the vision and foresight of J.J. de Gryse. The Laboratory of Insect Pathology was officially opened in 1950 in Sault Ste. Marie which was in the heart of the current spruce budworm infestation. The officer in charge was James MacBain Cameron and the scientists in the founding team were Gernot Bergold, Ted Bird who moved from Fredericton, Donald MacLeod, a mycologist from Nova Scotia, Thomas Angus and Arthur Heimpel as bacteriologists, and Hugh Thomson as a protozoologist.

Bain Cameron directed the Laboratory of Insect Pathology until his death in 1974. The name of the laboratory had been changed to the Insect Pathology Research Institute (IPRI) some years previously. Tom Angus was then Director for two years following three years as Associate Director. A major reorganisation occurred in the Canadian Forestry Service in 1976 when IPRI was amalgamated with its sister agency, the Chemical Control Research Institute (CCRI), in Ottawa, and the latter relocated in Sault Ste. Marie. The result of this union was named the Forest Pest Management Institute (FPMI) which has a staff of 82 and is directed by George Green. A new building was opened in 1985 and the original Laboratory of Insect Pathology fell victim to the wrecker's ball.

In the founding years, Gernot Bergold and Ted Bird were pioneers in the electron microscopy of insect viruses. Bergold published several landmark papers on the nature of polyhedrosis and granulosis viruses. Bergold left Sault Ste. Marie in 1960 and subsequently worked on human viruses and vaccines in Caracas, Venezuela. Ted Bird described the nuclear polyhedrosis virus of European pine sawfly and conducted the first aerial spray trial with a virus in Canada in 1950. Ed Steinhaus conducted the first ever aerial spray trial with a virus in California in 1949. Ted Bird retired in 1980 and died soon thereafter.

Peter Faulkner worked in Sault Ste. Marie in the 1950's and 1960's as a viral biochemist before joining the staff of Queen's University, Kingston where he is still researching insect viruses. Gordon Stairs, now at Ohio State University, worked for a few years in the 1960's on viruses of spruce budworm and forest tent caterpillar. Yoshiyuki Hayashi followed Peter Faulkner as a biochemist, but returned to Japan around 1970. Basil Arif followed Yosh Hayashi and is currently on staff working on the genetic manipulation and enhancement of spruce budworm viruses. John Cunningham was appointed to the Gordon Stairs position and he is active in field testing of viruses across Canada and registration of viral insecticides. Bill Kaupp is on the staff as a viral epizootiologist. The virology team has been successful in registering viruses for the control of redheaded pine sawfly and Douglas-fir tussock moth. A registration petition for European pine sawfly virus has been submitted and one on gypsy moth virus is being compiled. Other than *Bacillus thuringiensis* (*B.t.*) products, these are the only microbial agents currently registered in Canada.

Janina Krywienczyk was hired by Bergold to investigate the serology of insect viruses and published the first papers on this subject. She later worked on the serology of *Bacillus thuringiensis* strains, insect cell cultures and other organisms. Following her retirement, this position was filled by an immunochemist, Anthony Pang, who works on monoclonal antibodies to both *B.t.* and viruses.

Don MacLeod, who previously worked with fungal diseases of pea aphid in Nova Scotia, was the founder of the mycology project in Sault Ste. Marie. He was a taxonomist at heart and is internationally acclaimed for his studies on the genus *Entomophthora*. Don retired in 1982. Crossley Lougheed was on staff in the early 1960's working on fungal biochemistry. He left in 1964 and his position was filled by David Tyrrell who discovered a protoplast stage of some fungi when they were grown in cell cultures. Richard Soper worked on the epizootiological side of the mycology project during the late 1960's and early 1970's when he conducted computer model studies on cicadas and on woolly pine-needle aphid. On leaving FPMI, he headed for the Boyce Thompson Research Institute and is currently in administration with the USDA in Washington. David Perry joined the staff of FPMI about 1982 as a fungal epizootiologist with an interest in computer modelling. He moved to the Canadian Forestry Service laboratory in Quebec in 1985 and this year accepted a post with a biotechnology company in France.

Art Heimpel and Tom Angus, both founding members of IPRI, are remembered for their pioneering work with bacteria pathogenic to insects, and *B.t.* in particular. Tom Angus was the first to note the pathogenic nature of the *B.t.* crystal (proteinaceous parasporal inclusion) when this agent was fed to silkworms. Art Heimpel left Sault Ste. Marie around 1960 to become Director of the USDA Insect Laboratory at Beltsville, Maryland. Tom Angus retired in 1980 and, as an alderman, is still active in Sault Ste. Marie local politics. Paul Fast joined the bacteriology team in the 1960's as a biochemist who attempted to elucidate the toxic element of the protein crystal. Peter Lothy joined the staff of the bacteriology project in 1968 and worked mainly on the mode of action of *B.t.* He returned to his native Switzerland in 1973 and has continued research on *B.t.*

In the early 1980's, Paul Fast became involved in all facets of *B.t.* research ranging from deposit assessment and ultralow volume applications to biochemical studies. He formed a network called Biocide which involves two Canadian universities and Canada's National Research Council as well as FPMI. The aim of Biocide is to genetically engineer and develop a *B.t.* toxin which is highly effective against spruce budworm. Paul took early retirement in 1986, but is still retained as a Canadian Forestry Service (recently renamed Forestry Canada) consultant to Biocide. Currently the *B.t.* project, because that is what the FPMI bacteriology project has become, is headed by Kees van Frankenhuyzen who is involved in laboratory and field studies to optimize and expand the use of *B.t.* on several species of defoliating Lepidoptera.

The founding scientist of the protozoology project at IPRI was Hugh Thomson who published descriptive papers on microsporidia infecting spruce budworm and forest tent caterpillar. Following his death, this position was filled by Ren Ishihara who, like Yoshi Hayashi, later returned to his native Japan. Gary Wilson, who started work under Ted Bird, later trained as a protozoologist and took over this project in 1969. Gary furthered the work of Hugh Thomson on the microsporidian parasite, *Nosema fumiferanae*, which is exceedingly prevalent in aging populations of spruce budworm.

Insect cell culture goes hand in glove with insect pathology as does insect rearing. Silver Wyatt developed a medium for maintaining insect cultures in 1956. James Vaughn worked on insect cell culture at IPRI before moving to Beltsville in the mid 1960's. Following Art Heimpel's death, Jim became Director of the Beltsville Insect Pathology Laboratory and it is interesting to note that two directors of this establishment started their scientific careers in Sault Ste. Marie. Sardar Sohi took over the cell culture project in 1964. He has been successful in establishing many new cell lines for a variety of forest insect pests and has propagated several insect pathogens *in vitro*. A major breakthrough in insect rearing was made in 1966 when Arlene McMorran developed an artificial diet for spruce budworm.

This opened the door for greatly increased insect production, which in turn was used for such projects as virus production. Now many species are routinely reared on artificial diets.

When IPRI was established in 1950, it was decided to locate insect pathologists at key regional Canadian Forestry Service establishments to liaise and collaborate with scientists at Sault Ste. Marie. Murray Sager was first to hold this position in British Columbia and his death led to a vacancy filled by Oswald Morris in 1960. One of Ozzie's early assignments was an aerial application of *B.t.* on blackheaded budworm on the Queen Charlotte Islands in collaboration with Tom Angus. Ozzie worked with both *B.t.* and viruses and studied such western insect pests as Douglas-fir tussock moth and hemlock looper. He moved to Ottawa in 1970 to join the staff of CCRI where he worked on combinations of *B.t.* and low dosages of chemical pesticides. At FPMI in Sault Ste. Marie from 1978-82, he was involved in *B.t.* application technology. He then joined the staff of the Agriculture Canada laboratory in Winnipeg where he is currently working on *B.t.* and nematodes for control of agricultural insect pests. He is also involved in a Canadian International Development Agency project to control cotton pests in Egypt with *B.t.*

Murray Neilson was appointed regional insect pathologist at Fredericton, New Brunswick, where he published on viruses of winter moth and European spruce sawfly. He joined the ranks of management in the late 1960's and eventually became Director of the Maritimes Forestry Research Centre.

Vladimir Smirnoff joined the staff of the Quebec regional Canadian Forestry Service laboratory around 1960 and, in the beginning, worked with viruses, bacteria, protozoa, and fungi. He conducted aerial spray trials with Swaine's jack pine sawfly virus. However, in the 1970's and 1980's his efforts were directed to the operational use of *B.t.* on spruce budworm and his leadership and determination were instrumental in its widespread use first in Quebec and later in other Canadian provinces. Smirnoff's publication list is extensive and his enthusiasm and energy boundless; he retired in 1986. Jose Valero, an assistant to Smirnoff, is currently the only Canadian Forestry Service insect pathologist not located at Sault Ste. Marie.

Most of this history has been centered on insect pathology in the renamed Forestry Canada but Agriculture Canada also has an insect pathology program. Agriculture Canada's forte has been parasite research and a large laboratory was built after the Second World War at Belleville to house this endeavor which also supplied parasites to the Canadian Forestry Service. This facility was closed in 1972 which was a major blow to biocontrol research in Canada. Gordon Bucher came to Belleville in 1955 where he headed the program on insect pathology until 1972 when he transferred to Winnipeg. He is noted for his work on bacteria, particularly those infecting grasshoppers. In Winnipeg, Gord conducted research on bertha armyworm virus. He retired in 1982 and died soon thereafter. June Stephens worked with Gord Bucher in Belleville and studied insect immune systems using wax moth as a model. Now June Chadwick through marriage, she is on staff at Queen's University and continues her research on insect immunity.

Another notable insect pathologist with Agriculture Canada is Robert Jaques who has published on several insect pathogens including viruses, B.t. kurstaki, B.t. tenebrionis, fungi, nematodes, and microbial/chemical combinations. He started his career in Kentville, Nova Scotia, in 1954 where he initially worked on control of the apple sucker by naturally occurring E. sphaerosperma, a host-pathogen system studied earlier by Alan Dustan and Don MacLeod. He showed that certain fungicides applied to control plant pathogenic fungi did not affect this entomopathogen. Orchard tests in the late 1950's and early 1960's showed that B.t. controlled the winter moth and other foliage-feeding insects on apple. He found that the bacterium did not kill parasitic and predaceous arthropods and, therefore, mites and some other secondary pests were not a problem following application of B.t. whereas mites were a major problem if chemical insecticides were used. Bob began his studies on stability of viruses at Kentville and he continued this work when he transferred to the Canada Agriculture Research Station at Harrow, Ontario. His pioneer work on persistance and accumulation of viruses of the cabbage looper and imported cabbageworm in soil and on foliage contributed substantially to knowledge of epizootiology of insect viruses. In addition, Bob worked with viruses and B.t.k. against pests of cole crops, B.t.t. against the Colorado potato beetle, granulosis virus to control the codling moth, fungi to control whiteflies on greenhouse crops, and other host-pathogen systems.

There has been a considerable interest in entomopathogenic nematodes in Canada. Harold Welch must be regarded as the father of nematology in Canada with publications on nematodes infecting both terrestrial and aquatic insects dating back to 1958. His career started in Belleville with the Department of Agriculture in 1952. In 1965 he was appointed Head of the Department of Zoology, University of Manitoba, Winnipeg. He was the first person in Canada to test the practical use of nematodes for control of the Colorado potato beetle, lepidopterous pests on cole crops, and control of mosquito larvae. Harold has been on disability leave since 1985. Another nematologist who started his career in Belleville is John Webster who studied host parasite relationships and is currently at Simon Fraser

University in British Columbia. Jean Finney, now Jean Finney-Crawley, works at Memorial University in Newfoundland. She is currently involved in both the production and screening of mermithid nematodes for blackfly and mosquito control, and steinernematid nematodes for control of agricultural and some forestry pests.

Although several microbial agents are used in Canada, nematodes are the only ones produced commercially in Canada. All *B.t.* is imported and viruses are presently produced only by government agencies. Phero Tech Inc., a British Columbia company which markets pheromone traps, has expanded into nematodes and is producing *Heterorhabditis heliothidis* and *Steinernema feltiae*. The former is excellent for controlling house flies in manure heaps and black vine weevil in greenhouses; Phero Tech staff are investigating other uses. Byosis, a nematode-producing company located at Palo Alto, California, has opened a branch in an Alberta industrial development park and commenced massive in vitro production of steinernematid nematodes. A third company interested in nematodes in Canada is called Philobios.

Several scientists who have worked on pathogens of agricultural insect pests are located at the Agriculture Canada Research Station in Saskatoon. Alwyn Even has researched the use of the microsporidian, *Nosema locustae*, for grasshopper control and is involved in an attempt to register this pathogen in Canada. Alfred Arthur and Martin Erlandson have published on various aspects of insect pathology. David Johnson at the Agriculture Canada laboratory at Lethbridge is also interested in the development of *N. locustae* for grasshopper control in Canada.

Formal instruction on insect pathology has not been offered at a Canadian university and Canadians have found it necessary to travel to the United States to obtain undergraduate training. However, opportunities for graduate work do occur at some Canadian universities, and several internationally acclaimed insect pathologists have been faculty members or are currently on staff at various Canadian universities.

Edouard Kurstak, after whom *B.t.* variety *kurstaki* is named, is a professor in the Faculty of Medicine, University of Montreal. His team published several papers on densonucleosis viruses of insects. More recently Prof. Kurstak has edited several books on viruses generally and organised workshops on comparative virology. Serge Belloncik at the Armand Frappier Institut in Quebec and his team have worked with viruses, protozoa, and fungal diseases of mosquitos, and viruses of agricultural insect pests. Don Stoltz at Dalhousie University, in Halifax, Nova Scotia has made a name in the rather obscure field of viruses associated with parasitic Hymenoptera.

In the 1950's, Chris Hannay, a bacterial cytologist and electron microscopist at the University of Western Ontario in London, was interested in the morphology of crystalliferous bacteria. His shadowed electron micrograph of a *B.t.* crystal showing its fine structure must be one of the most reproduced pictures in the whole field of insect pathology! Russ Zacharuk at the University of Regina published an elegant series of papers in the early 1970's on electron microscopic observations of the fungus *Metarhizium anisopliae* penetrating the cuticle of wireworms and invading the host. Peter Lee at Carleton University, Ottawa published several papers on bee viruses in the 1960's and has taken an interest in the electron microscopy of other groups of insect viruses. Currently, Peter Faulkner is active in the field of molecular biology of insect viruses and June Chadwick is one of the few researchers worldwide working on insect immunity. They are both located at Queen's University, Kingston, and both have been mentioned earlier in this account.

An intriguing episode in the history of insect pathology in Canada concerns the Unit of Vector Pathology, funded by WHO and located at Memorial University, St. John's, Newfoundland. It was founded in the early 1970's by Prof. Marshall Laird who had an innovative approach to biological control of blackflies and mosquitos. In the beginning, all potential pathogens known at that time, microsporidia, fungi, and mermithid nematodes were studied. Considerable emphasis was placed on *in vitro* production of nematodes. About 1978, *Bacillus thuringiensis* var. *israelensis* became commercially available and considerable resources were expended on the use of this agent both in a WHO program to control onchocerciasis (river blindness), caused by a parasite transmitted by blackflies in Upper Volta in West Africa and in evaluating its impact on mosquitos and blackflies inhabiting the cold waters of Newfoundland and Labrador.

Unfortunately, the unit was disbanded in 1984 and Marshall Laird returned to his native New Zealand. Other staff members of the unit were Richard Nolan who worked on fungi, Roger Gordon and Charles Bailey who worked on nematodes, Joseph Mokry and Murray Colba were ecologists, Albert Undeen did the first basic work with *B.t.i.*, Jean Finney-Crawley studied the effect of cold weather conditions on nematodes and *B.t.* Marshall Laird, a renowned world traveller, frequently visited Upper Volta; Murray Colba and Al Undeen also ventured to this obscure corner of the globe to conduct field trials with *B.t.i.* There was close collaboration between this Unit and a USDA station at Lake Charles, Louisiana, which cultured a mosquito parasitic mermithid nematode *in vitro*. This nematode was also used in West Africa against blackflies.

While on the topic of biting flies, it is interesting to note that the City of Winnipeg is a leader in the use of *B.t.i* for mosquito control. Its use was encouraged by Mary Galloway, Director of the Biting Fly Research Centre, and the city entomologist, Roy Ellis who has recently retired to become a consultant and has been replaced by Randy Gadawsky.

To conclude this somewhat anecdotal review of insect pathology in Canada, it must be pointed out that, unlike the USA with Ed Steinhaus, there is no single person who can be identified as the father of Canadian insect pathology but rather several far-sighted individuals who laid the foundations of this science in Canada. Canadians are probably the best practitioners of microbial control worldwide, particularly in forestry. Canada is the largest user of *B.t.* var. *kurstaki* and vast areas are aerially sprayed annually to control such defoliating Lepidoptera as spruce budworm and gypsy moth. There are currently 22 *B.t.* var. *kurstaki* products registered in Canada and 11 *B.t.* var. *israelensis* products. The latter bacterium is widely used by municipalities for mosquito control and by private contractors in recreational areas for blackfly control. Of the two viruses registered in Canada, redheaded pine sawfly virus is used routinely to combat this pest in plantations in Ontario and Quebec. A Douglas-fir tussock moth outbreak is expected in British Columbia in 1990 and B.C. Forest Service has adequate supplies of virus (both Canadian and American products) on hand to combat this cyclical pest.

Entomopathogens are not used as extensively in agriculture in Canada as they are in forestry, not only because registration for use on food products is more difficult than for use on forests, but also because individual growers determine their own procedure to combat pest problems rather than being strictly controlled by government agencies who are responsible for protection of large areas of forest. An attempt is being made to register *Nosema locustae* for grasshopper control in Canada; this may be successful in the near future. *B.t.* var. *kurstaki* is registered and available for use against a number of agricultural pests. A major use is the control of lepidopterous pests of cole crops but, although it is very effective and environmentally safe, the bacterium is used presently on only about half of this crop. Registration of codling moth granulosis virus in Canada is being pursued. If this virus is available, it will certainly be in demand by organic growers who need to control this pest by biological means in order to market pesticide-free apples at a reasonable price. Similarly, fungi have potential for controlling whitefly on greenhouse-grown vegetables making the insecticide-free production of these crops a feasible venture.

As we approach the 21st century, the next chapter in insect pathology will probably belong to our geneticists and molecular biologists who will modify, enhance, and improve existing strains of pathogens and develop a second generation of microbial pesticides which will, we hope, be more effective and cost less than currently available products.

Acknowledgement

I wish to thank the following people who supplied historical information and/or reviewed this manuscript and offered helpful advice and criticism; in alphabetical order, Dr. T.A. Angus, Mrs. M.B.E. Cunningham, Dr. J. Finney-Crawley, Dr. G.W. Green, Mrs. K. Jamieson, Dr. R.P. Jaques, Dr. W.J. Kaupp, Dr. O.N. Morris, Dr. D. Tyrrell, and Dr. G. Wilson.

Selected Reading

Angus, T.A. 1964. Canadian participation in insect pathology. Can. Entomol. 96: 231-241.

Kelleher, J.S. and M.A. Hulme (eds.). 1984. Biological Control Programmes against Insects and Weeds in Canada 1969-1980. Commonw. Agric. Bureaux, Slough, U.K. 410 pp.

Morris, O.N., J.C. Cunningham, J.R. Finney-Crawley, R.P. Jaques and G. Kinoshita. 1986. Microbial insecticides in Canada: their registration and use in agriculture, forestry and public and animal health. Rept. Special Committee of Sci. Policy Committee, Entomol. Soc. Canada. 43 pp.

Prebble, M.L. (ed.). 1975. Aerial Control of Forest Insects in Canada. Dept. of Environ., Ottawa, Canada. 330 pp.

Steinhaus, E.A. 1975. Disease in a Minor Chord. Ohio State University Press, Columbus. 488 pp.

Chapter 4

Insects Around the Home

THE FAR SIDE By GARY LARSON

The last thing a fly ever sees

War and Peace
in Wardrobe and Pantry

J. Richard Gorham

Every time I think of those stories in the Bible about the Egyptians storing up great quantities of cereal grains to see them through times of famine, I wonder how they managed to keep insect and mite pests out of the granaries. Apparently they achieved some success in this effort, but no system of storage, whether ancient or modern, was or is ideal in every respect.

Whenever a person sets aside something in storage — it could be a cup of cornmeal on the pantry shelf, or two million bushels of wheat in a Kansas elevator, or it could be a neglected woolen mitten lying on the closet floor, or a warehouse full of Kashmiri rugs — there's a good chance that some kind of storage pest will discover it and convert some part of it to its own use.

The arthropods I'm alluding to here are mainly mites, moths, and beetles, but some belong to other groups as well. There are thirty or so really important ones, but a more inclusive list could easily run to 200 species. This is a very diverse collection of animals in terms of habits and feeding preferences. Some feed largely or exclusively on plant materials; others eat only matter of animal origin. At least three associated groups exist: predators, parasites, and scavengers. The predators eat the insects and mites that come to the stored products. The parasites, mostly tiny wasps, also eat the pests but they do it much more slowly than the predators. Then the scavengers come along and pick through the leftovers.

Some of the plant-feeding pests prefer whole seeds of cereal grains or legumes. Their larvae feed quite unseen within the seeds. Others require broken seeds or even small particles such as cornmeal, or flour, or breakfast cereals. Some are content only on certain kinds of foods — bean weevils on beans, rice weevils on rice, and others will eat a wide variety of plant materials.

The pests that rely on foods (from their point of view, at least) of animal origin are represented mainly by the dermestid beetles and the clothes moth and its relatives. The life styles of the dozen or so very common dermestid species vary widely. Some prefer animal flesh that has dried on the bone. Others go for wool clothing. Still others eat fur or feathers or dead insects. And some species will feed on virtually any dry-matter of animal origin (powdered milk, for example).

A few species of dermestids have gone over to the other side and now feed mainly on plant materials. The most notorious of these is the khapra beetle (known to entomologists as *Trogoderma granarium*), a native of southern Asia. From time to time the khapra beetle reaches the United States on imported goods. When that happens, it triggers a major effort to eradicate the beetle before it becomes established.

The "peace" that we seek to make with these intruders on our food and clothing is actually more of a truce as it comes only after we have engaged in "open warfare" against our enemies, and it is maintained only by constant vigilance on our part. We can rest assured that our insect competitors are constantly "vigilant" for any opportunity to break through our defenses.

In most cases, our tactics are similar whether we are trying to protect our cup of cornmeal or our warehouse full of Kashmiri rugs.

Sometimes we use exclusion. If, for example, we put our cornmeal in a jar with a tight lid instead of in an open cup on the pantry shelf, mites and insects won't be able to get to it. The ancient Egyptians sometimes sealed their barley in mud-brick granaries. This tended to keep out the pests but did little to stop the proliferation of those already inside. We face the same problem today, and that's why we recommend sanitation (good housekeeping) as an aid in the prevention of pest problems. We want to clean up and clean out storage areas to get rid of both pests and potential food for pests — so that newly introduced products won't become infested.

Sometimes we purposely do such a good job of exclusion that we make the storage area airtight. This not only keeps pests out but as the oxygen gets used up, the pests eventually die for lack of oxygen. We can go a step further when the storage area is airtight and deliberately introduce inert gases such as carbon dioxide or nitrogen to displace the oxygen and consequently kill the pests.

Now suppose we put our cup of cornmeal in the refrigerator instead of on the pantry shelf. This won't kill any insects or mites that happen to be present, but it "sure does slow 'em down!" They can't eat and they can't reproduce. This same principle has been tried on a much larger scale. One technique is to blow cool air through grain stores to slow down pest activity. Going a step further, the Israelis have constructed enormous silo-type granaries through which they pump refrigerated air. The technique works well but requires a considerable investment of energy

When all else fails, we resort to the judicious use of chemicals. If circumstances permit, we can cover the infested product and pump in a fumigant gas. When held for sufficient time at the right temperature and at the right concentration, the gas kills all insects and mites in the product. But the effect of the fumigant doesn't last, so as soon as the fumigation job is complete, the product is immediately at risk of reinfestation.

Another chemical approach is to clear everything out of the pantry and spray its walls and shelves with an approved residual insecticide. This residual toxicant will kill any insects that happen to touch it, but it will have no effect on insects that are already inside boxes of food and are content to stay there. So, along with the use of chemicals, we need to inspect the products and dispose of the ones that are infested. This procedure of visual inspection and residual application works just as well for a food warehouse as it does for the kitchen pantry.

A final technique in our war with storage pests is the use of repellent materials. This technique has not been used much around food but it is widely and successfully used against clothes moths and carpet beetles. Mothballs repel the bugs that eat our woolen mittens, but they can be successfully used only in an enclosed space with very limited air circulation. Now this method presupposes that our mittens are clean and free of bugs before we store them. If our mittens are dirty and/or infested, the protective effect of the mothballs is greatly diminished. These criteria for safe storage apply to all materials composed at least in part of fur, hair, wools, leather, or feathers. The only really safe way to store small items composed of any of these animal materials is to put them in a freezer.

As you might imagine, a problem that touches not just the big warehouses and grain silos, but one that has reached into the pantry of every household in the world at one time or another has attracted the attention of at least a few entomologists. These men and women of science have studied and are continuing to study both the biology of the pests and the techniques for reducing the negative impact of these pests on the world's food supplies. A lot has been learned. Entomologists have, for example, discovered chemicals called pheromones that the insects and mites themselves use to communicate with each other. These same chemicals are now being used against the pests, making them less destructive. Another example: Entomologists have worked out a system to fumigate export grain ships while they are in transit to foreign ports. The fumigant prevents the pests from damaging the grain during long sea voyages.

But there is still much to learn about these competitors that keep food from reaching our mouths, that destroy the fur and woolen clothes that we need to keep warm, and that ruin the woolen carpets and tapestries that make our homes comfortable and beautiful. Good minds are invited to join in the fun of figuring out ways to protect the food and fiber of today's and tomorrow's world.

Suggested Reading

Brunner, H. 1987. Cold preservation of grain. In Proceedings of the Fourth International Working Conference on Stored-Product Protection (Tel Aviv, 1986), pp. 219-229, ed. by E. Donahaye and S. Navarro.

Gorham, J. R. (ed.). 1991. Ecology and Management of Food-Industry Pests. Association of Official Analytical Chemists, Arlington, Virginia.

Gorham, J. R. (ed.). 1991. Insect and Mite Pests in Food: An Illustrated Key. Agriculture Handbook 655. U. S. Government Printing Office, Washington, D.C.

Levinon, H. Z., and A. R. Levinson. 1985. Storage and insect species of stored grain and tombs in ancient Egypt. Journal of Applied Entomology 100(4)321-339.

Reichmuth, C. 1987. Low oxygen content to control stored product insects. *In* Proceedings of the Fourth International Working on Stored-Product Protection (Tel Aviv, 1986) pp. 194-207 ed. by E. Donahaye and S. Navarro.

Urban Entomology — The Sound of One Hand Clapping?

Richard Carr

The elusive urban entomologist

By now you have guessed that entomology is a subdiscipline of the scientific study of animals (zoology), which is itself a subset of the study of all life forms (biology). Entomology confines its scientific investigations to one group of animals, the six-legged animals (hexapods) — the insects.

Having said this much — you should know that within entomology there are also sub-disciplines or more accurately, special areas of interest. For example, an "agricultural entomologist" will generally confine his or her studies to those insects that live in and/or feed upon the cultivated organisms that we utilize as human food sources, and usually lump together under the word, crops. Yet, not all of these "six-legged science" endeavors are so easily described. Who are the urban entomologists and what do they do?

Urban entomology is one of those difficult to define areas of special interest to a rather small group of entomologists. However, the services of these scientists are increasingly in demand by some informed citizens.

Students of the eastern "Way of Liberation" that the Japanese call, Zen, are often given mind numbing conundrums to solve. These exercises, in attempting to solve the unsolvable, are called, in Japanese, "koans" (literally, "a legal case"). Such a koan is, "Show me the sound of one hand clapping." One of the points of such activities is to help the student of Zen to discover the foolishness of seeking his or her real self in the abstract terms of conventional thinking. "Show me the shape or substance of urban entomology" might well serve as a Zen koan.

To better understand who urban entomologists are and the areas in which they work, it would be best if you try to suspend your conventional, one-thought-after-another, abstract way of thinking — at least, while you read this brief essay. Just read and let the information flow into your mind. If thoughts arise, pay no attention to them. You will no doubt recall them later and then, if you desire, you can grasp at them in the manner we Westerners have been taught to "think."

Now — you are in your cozy, high-rise condominium located along the lake shore on the near-north side of Chicago. You are sitting in your favorite chair and reading this little book on professional entomology. You have on the end-table at chair-side a cup of hot tea, and you are relaxed. You reach for the tea. As your hand moves toward the cup, your glance follows the movement of your hand —. SUDDENLY, you are startled by what appears to be a small creature skittering across the table-top, headed for your tea cup!

With no more information than — this creature is, in fact, an insect, allows me to ask you a question, "Would this scenario fit into the province of urban entomology?" This is not a question intended to trick you —, 'though it may smack a little of Zen. Don't think! ANSWER! Yes or no?

Probably most of you would answer yes. Why? Because I predisposed your mind to a "city" situation — or an "urban" setting. Although this scenario would indeed be in the province of the urban entomologist, I could have placed you in Tucson, Arizona, covered with sun tan oil, lying on a beach towel on the grass, in the backyard of your threebedroom, spanish-adobe home. SUDDENLY, a small "bug" (in fact, an insect) leaps from the grass and lands in the midst of your oil-coated stomach! This scenario, however, would most likely be in the province of the "turf-grass entomologist," a sometimes urban entomologist.

Entomologists (as with other scientists) can be a rather clannish group of people, given overly much to doing their own-thing, and in the process, becoming very territorial or excessively protective of their tightly focused, special interests. Urban entomologists have not escaped this characteristic and narrow behavior. Given that this is the case, you would think that these scientists would have rigidly defined, "urban entomology" —, the better to protect it from other grasping, bug-scientists. Not so! In fact, no official or generally agreed upon definition of who or what an urban entomologist is or what he or she should study exist — ("...the sound of one hand clapping"). But, I will now do my best to tell you who I think such a scientist may be, and why he or she might be of practical service and of real benefit to you.

A Little History

As various areas of special interests in the study of insects arose over the last two or three centuries, some areas developed and were based upon the identification, collection, display, and/or the study of the biology and behavior of a particular group of insects, such as the butterflies, or beetles. Some of these entomologists narrowed their interests even further to one genus of butterflies, or other insect group. These men and women often sought knowledge of insects because of the joy of discovery and the adventure of navigating within a hitherto unknown realm of nature.

Since the turn of this century, many entomologists are concentrating their studies upon only those insects that cause damage to crops, contaminate food, infest domestic animals, or — in order to control other pests — those insects that prey upon one another or other biological organisms. Other entomologists are meeting the challenges of studying the insects which can transmit diseases that are debilitating or fatal to humans. These latter entomologists are closely allied with medical professionals, and they prefer to call their special area of entomological study, "medical entomology."

The entomologists who study the identification and biology of insects, primarily to expand human knowledge, often refer to themselves as, "basic studies entomologists" (or simply "basic entomologists"). Whereas, the group who study insects in order to find ways to lessen the impact of "pest" insects upon human society are called, "applied studies entomologists" (or "applied entomologists"). The urban entomologists, which have grown in numbers especially since the Second World War, generally view themselves as applied entomologists.

What Some Urban Entomologists Do

Urban entomologists can be found as employees of private industry, state universities, private colleges, federal and state agencies, members of the military and other uniformed service organizations, and as private practitioners (consulting entomologists). It is the urban entomologists at the state universities (particularly Cooperative Extension Entomologists), private industry urban entomologists (particularly those who work for urban pest management firms — often referred to, unfortunately, as "exterminators"), and the consulting urban entomologists that may be of most immediate value to you as a consumer of goods and services in urbanized North America.

For example, you rise late on Sunday morning and open the pancake flour. You notice that the surface of the flour seems to be animated by several small, brownish colored "bugs." After looking around your kitchen you also find some of these flour-creatures in your cupboards and on the kitchen counter-tops. What should you do?

There are many possibilities to be considered to solve this pest problem. You could ignore the problem, but within a few months you will probably no longer be able to ignore the burgeoning population of little bugs. You could go to a general merchandise store (today, we call them "discount" stores) and purchase some insecticide (you are assuming the pest is an insect), but you sometimes have allergic responses to some chemical odors. Still — you can call an urban pest management firm to solve the problem. Perhaps they will use a non-chemical insect control strategy. You might also call upon a consulting entomologist, although their fees are for professional services and may be more than you want to pay — at least until you better understand the complexity of the pest problem.

But, how can you gain a better understanding of this situation if you don't call on some of these professionals? Don't despair, you still have other choices!

If you live in or near to a large city, there is probably a local office of the state university's Cooperative Extension Service not far from you. Call them and request that you be allowed to speak with someone knowledgeable about urban, household, or indoor pests. Your call will probably then be referred to an urban entomologist, and after explaining the situation you may discover that the solution to your problem is simple, low cost, and safer than you imagine.

Another very valuable area where an urban entomologist can save you money and a lot of heart-ache is when you decide to purchase a new home. In many states, and for all VA-FHA federally financed home purchases, a wood-destroying organisms or wood-destroying insects inspection of the house and issuance of a report of the inspection findings is required before the sale/purchase transaction can be completed. In many of these states the burden of acquiring and paying for this inspection is on the seller. However, in the last few years, some states are requesting that the buyer assume this responsibility. Actually, it is to the buyer's advantage to request his or her own inspection be done — whatever the federal, state, or local laws may require (reason: some persons still exchange goods and services on the basis of "caveat emptor" (let the buyer beware).

Would an urban entomologist be the appropriate professional to call to have such a "structural inspection" made and a report written? Generally, yes. However, it is also appropriate to seek out those urban entomologists who have vested interests in doing the best job for you.

Many urban pest management firms have staff entomologists who can do these inspections. A growing number of private practitioner or consulting entomologists will do structural inspections. An extension, government, or any other public agency urban entomologist is generally not permitted to do these private transaction inspections.

Since the purchase of a house in today's marketplace can range into the hundreds of thousands of dollars, it is best to check the credentials of any urban entomologist called on to do the inspection, no matter who he or she may work for. To do this — you could check this persons name with the Entomological Society of America in Lanham, Maryland. This organization of professional entomologists has within its structure a registry of Registered Professional Entomologists (R.P.E's.). These men and women have met professional standards beyond their graduate degrees. The registry divides its scientists into seventeen entomological specialties one of which is the specialty of "Urban/Industrial Entomology" — Just the list you are looking for!

How do government urban entomologists serve the public? Some do basic research; others work on the development of new pest management strategies. Those that work in the various branches of the military, for example, often do double-duty as medical and urban entomologists. Double-duty is also the role played by many public health entomologists, such as those employed by the Food and Drug Administration. These FDA scientists who help to keep our food wholesome also must often be accomplished urban entomologists.

Urban entomologists may be employed by commercial service industries. The urban pest management firm called on to inspect the new house purchase mentioned earlier is an example of this type of urban entomologist. Urban entomologists are also employed by various types of manufacturing or industrial firms. These scientists may be responsible for assuring that insect and sometimes other pests are not processed and packaged with our food. In other industries they might assist in the research and development of new pest management products from chemicals to the new biotechnology pest control concepts and agents.

Finally, some of the extension urban entomologists that assisted in controlling the pancake flour-creatures in the example described previously also do research, most often applied. Others may teach, and their teaching is frequently aimed at educating technologists of various urban products manufacturing or services industries. Universities also employ other urban entomologists to do basic research and formal student teaching. These men and women have no extension responsibilities, yet their work lays the knowledge base for the technology development of the applied urban entomologists. They may also provide basic education for other urban studies professionals from structural engineers to social workers, as well as educating a new generation of entomologists.

In a nutshell, urban entomologists are directly involved in the protection of post-harvest food, wood-bearing shelter, animal or plant derived fibers (for example, cloth made of wool, cotton, silk, etc.), the protection of domestic animals, and in concert with medical entomologists — human health. How important are these entomologists?

How much value would you place on the food you eat, the roof over your head, the clothes you wear, or your own life?

By now you should be getting the idea that urban entomology is a very broad field. Entomological work done primarily in an urban setting, and on insects that occur most often in urbanized situations are two of the elements that all urban entomologist have in common. Their scientific efforts affect almost everything that occurs or exists in our rapidly growing urban communities. These professionally trained men and women are critically important to our wellbeing and to the enhancement of our standard of life on planet Earth.

Before you read this article you may not have known who or what an entomologist might be; now you know that entomologists are scientists who study insects. As for urban entomologists, well — I still have not rigidly defined them for you or locked them up in an "official" description. I have tried to point directly at what some of these scientists do, and at the importance their work has for each of us as city-dwellers.

The Sound of One Hand Clapping?

Foolishness indeed! One hand clapping does not produce sound, but urban entomology does. It is the sound of professional activity, the sound of serving, of making life somewhat better for us than it might otherwise be.

In conclusion, allow me to answer one of those questions that you just couldn't stop yourself from grasping while you were reading this article. Question: Why would anyone want to become an urban entomologist? My personal answer to that question (since I am an urban entomologist) is summed up in the little portion of Zen verse below that describes the awakening from a Winter's sleep of the box-elder bug, a rather harmless "pest" commonly discovered during Spring cleaning in many North American homes:

"Sunlight kindles fire in the heart of this little charcoal
creature on my window sill.
Spring comes — just so."

Insects and Pest Control

George W. Rambo

Humans have lived with insects since time began. As we developed our needs increased and we came into conflict with insects and related arthropods. In these cases, they became pests.

In perspective, pests are animals or plants (here we are dealing with insects), living their natural existence. It is man who has intruded. Our affluence and mobility within our society has placed us in many situations where insects become pests. In the last 100 years this has become increasingly important. Economically, aesthetically, and for health reasons, we do not, or cannot, tolerate certain insects to impact on our society. In agriculture, insects can destroy crops in the field and harvests in storage. With the present demands on modern agriculture this is unacceptable. Man used natural "pesticides" against these insects in the past, such as arsenic, nicotine, and other organic materials. Later, synthetic chemicals took their place. These advances, and the integration of pest management concepts of scouting and biological control, have enabled us to lead the world in agricultural production. The continued development of "new pesticides" to combat resistant insects has lead the industry into the use of "natural chemicals" (by these we mean pheromones which attract insects for trapping and monitoring, and insect growth regulators (IGRs) — juvenile hormones and chitin inhibitors). IGRs have been used in stored product insect control and some other interior pest control problems. These "pesticides" sterilize adults and/or disrupt the insect's life style.

In agriculture, the use of "natural chemicals" is limited. Pheromones are used largely for monitoring and then to determine if other measures should be taken. Examples of this are fruit fly problems in California and Florida, and gypsy moth spread in the Central Atlantic states. This is still a developing science.

The use of predators has been explored. Parasitic wasps, mites, nematodes, and even fish are organisms that have been tested. Several tiny wasp species will attack the tomato hornworm, Mexican bean beetle, and other pest insects. Mites have also been isolated that attack some species of ants and termites. Nematodes have shown effectiveness against corn earworms and grubs that damage turf. Last but not least, fish have been utilized to feed on mosquito larvae. These are a few examples of solutions to pest problems being explored in agriculture and some urban arenas. As a result of modern pest control, we experience food losses of only 9% in the U.S. compared to 40%-50% in underdeveloped countries.

While agriculture was developing to produce food for millions of people, the structure of society has changed from one of agrarian and rural to highly urbanized. As more people moved into the cities and affluence grew, so did our concern for our health and the belongings around us.

In the America of yesterday, we lived and shared our homes with many pests and lived with the knowledge that deadly or debilitating disease organisms — many carried or transmitted by common insects — could strike us or our families at any time.

● In the early 20th century, there were sporadic outbreaks of bubonic plague in the United States, the same "Black Death" that killed 25 million people in Europe in the 14th century. The cause: infected fleas from rodents.

● In the 1930s, there were 6-7 million cases of malaria annually in the United States. The cause: the malaria parasite transmitted to humans by the *Anopheles* mosquito.

Now, we live in a country where plagues and epidemics are a vague memory. No one worries at the thought of Americans getting malaria, yellow fever, or dengue fever. Few Americans have seen typhus, typhoid, cholera, bubonic plague, or even dysentery. These diseases were once common afflictions in America, transmitted by pests to humans.

As society developed and sanitation and housing improved, pest problems decreased. Some of this decrease was attributed to the use of pesticides. During World War II, the chemical industry developed several powerful chemicals that were found to have a wide spectrum of activity against pests. In many situations, these pesticides, such as DDT, saved millions of lives in Europe and war-torn countries, as well as in the U.S.A.

As the chemical industry grew, newer insecticides were developed to give people the opportunity to increase the health and well-being within their world. But insect pests are still with us today and the disease organisms they can spread still abound.

Even in the 1990s, man is still plagued by a variety of serious pests:

● Body lice can carry the organisms for epidemic typhus fever and relapsing fever. Body lice have become a more important pest in modern America with homeless people. Head lice are reported every year in many school systems.

● Ticks, an insect-related arthropod, transmit the organisms for relapsing fever, Rocky Mountain spotted fever, and Lyme disease to man, and Texas fever to cattle. A sometimes fatal disease, Rocky Mountain spotted fever occurs through much of the United States.

Over 23,000 cases of Lyme disease, a potentially debilitating illness, have been reported in 42 states since 1980.

● The house fly harbors more than 100 kinds of pathogenic organisms and may transmit more than 65 human and animal disease organisms.

● Cockroaches transmit the organisms for such diseases as food poisoning, cholera, dysentery and perhaps typhoid. In one study, 89% of the cockroaches sampled carried at least three kinds of disease-causing bacteria on their bodies. Now, more evidence has indicated allergic reactions to insects, especially the cockroach.

● Fleas directly affect the health and well-being of pets and their owners. They can cause dermatitis in pets and spread parasitic worms.

Many other insects additionally plague man with bites, stings, or through their destructive habits.

Disease and health concerns from insects is a real "fact of life" we can, and have, dealt with well.

● We are kept free of many pests and pest-borne diseases not because of great advances in modern medicine, but because of modern pest control and improved sanitation. Drugs are no substitute for the elimination of the insect and/or rodent vector for disease organisms. In addition, we no longer have to share our homes with cockroaches, bedbugs, ants, moths, and silverfish. The pest control industry is playing a major role in helping us live healthier and more comfortably than ever before.

In many ways, we have become complacent in our acceptance of this situation. We also feel in many ways that insects have been put here to annoy us or plague us, rather than being a naturally occurring fact of life. Man still has problems accepting insects interfering with his picnic, boating, camping, or other outdoor activities. However, in these incidences, we are the intruder.

References - Historical Perspective

Link, V. B. 1955. A History of Plague in the United States, Public Health Monograph No. 26, U.S. Department of Health, Education and Welfare, Public Health Service.

James, M. T. & R. Hardwood. 1969. Herms' Medical Entomology, 6th ed. Macmillan Publishing Co., Inc., New York, 484 pp.

Seligmann, J., M. Hager, L. Drew, T. Padgett, N. De La Pena, K. Robins, & T. Clifton. Newsweek. May 22, 1989. Tiny tick, big worry.

Ebeling, W. 1975. Urban Entomology. University of California, Division of Agricultural Sciences, 695 pp.

Alcamao, I. E. & A. Frishman. March/April 1980. The microbial flora of field collected cockroaches and other arthropods. J. Environmental Health.

Pinto, L. J. 1981. The Structural Pest Control Industry: Description and Impact on the Nation. National Pest Control Association, 36 pp.

Kang, B. 1990. Impact of cockroach allergens on humans. Albert B. Chandler Medical Center, In Proceedings of the National Urban Entomology Conference, 11 pp.

Termite and Beetle Research at the Wood Products Insect Research Unit at Gulfport, Mississippi

Joe K. Mauldin

The current mission of the Termite and Beetle Research Project at the Gulfport laboratory is to develop chemical, biological, or physical methods of controlling, or preventing, damage by termites and wood-destroying beetles to wood in storage and in use. With some temporary additions and deletions, this mission has been the project goal since this research was begun in the late 1930s.

Termites and wood-destroying beetles occur throughout the United States of America and its possessions. Termites damage a wide variety of building components and contents in addition to a number of other products such as poles, pilings, and underground electric cable insulation. Damage by wood-destroying beetles is usually confined to building components and manufactured articles. Although the total economic impact of either insect group in not known, termite damage to buildings in this country was estimated at $750 million in the year 1980, a significant drain on the national economy and forest resources. Damage caused by the Formosan subterranean termite in Hawaii, and in some southern states, has increased the loss significantly. In Hawaii alone, this insect causes an estimated annual loss of about $60 million. The cost of damage by wood-destroying beetles may exceed $50 million annually.

In 1938, Mr. Harmon R. Johnston was assigned as the first entomologist in this project located at the Harrison Experimental Forest about 20 miles north of Gulfport, Mississippi, in the DeSoto National Forest. His research assignment dealt with biology and control of subterranean termites and ambrosia beetles. The latter insects attack green logs and lumber and are a problem at numerous mills across the south. Johnston was under the direction of Dr. T. E. Snyder, an internationally known and respected termite specialist, who was, in 1934, the first entomologist assigned to the Southern Forest Experiment Station in New Orleans, Louisiana.

In 1945, the U.S. Army Corps of Engineers provided funds to greatly expand the research on control of subterranean termites. About this time, Snyder transferred to Washington, D.C., and his successor, Dr. Joseph R. Kowal, was stationed at Gulfport.

In 1946, three additional entomologists, Mr. Sam Dews, Mr. J. Vaughn, and Mr. Herb Secrest, were hired, making a total of five professional entomologists assigned to the Gulfport project at that time. Although the project was mainly involved in wood products insect research, some time was devoted to studying and making recommendations concerning control of insects attacking forest trees.

Two new entomologists, Mr. John F. Coyne and Mr. Robert Morris, were hired in 1948 to replace Dews and Vaughn and the research was expanded to include sawflies, ips bark beetles, southern pine beetles, and black turpentine beetles. The project was then responsible for all work on forest insects for the Southern Forest Experiment Station region (Alabama, Arkansas, East Oklahoma, East Texas, Louisiana, Mississippi, Puerto Rico, Tennessee and the Virgin Islands), and for all work in the United States on insects attacking wood products.

During the period 1946 to 1958, several changes in personnel were made. Johnston was appointed project leader in 1954 when Kowal transferred to Asheville, North Carolina. Morris was reassigned to the Panama Canal Zone in 1952 to begin accelerated testing of new insecticides against termites, and two more entomologists, Mr. Virgil K. Smith and Mr. Raymond H. Beal, were hired in 1956 and 1958, respectively. The number of scientists engaged in this research remained about the same — three or four. By 1961, most research on insects that attack living trees was transferred to other locations. However, wood products insect research remained at Gulfport.

Studies were initiated during the period from 1946 to 1958 that resulted in improved control methods for subterranean termites. Some of the studies are still in progress in the Harrison Experimental Forest. These are the studies that proved the effectiveness of the chlorinated hydrocarbon insecticides — aldrin, chlordane, dieldrin, and heptachlor — for subterranean termite prevention and control. For over 30 years these chemicals were successfully used to protect wooden buildings and other wood products in the United States. Control methods for ambrosia beetles and bark beetles in logs and pulpwood were also developed. Benzene hexachloride was proven effective against ambrosia beetles, bark beetles, and southern pine beetles. Lindane (the gamma isomer of BHC) is still being used to protect green logs from ambrosia beetles.

Snyder's work resulted in a series of definitive publications of worldwide importance that were titled "Catalog of the Termites (Isoptera) of the World."

In 1965, considerable pesticide act monies were allotted to the project to expand research on wood products insects. As a result, a chemist, Ms. Fairie Lyn Carter, was hired in 1965; three entomologists, Dr. Richard V. Smythe, Dr. Joe K. Mauldin, and Mr. J. P. Secrest, were hired in 1966; and, in 1967, another entomologist, Mr. Lonnie H. Williams, was hired to fill the position vacated by Secrest. The number of scientists at this time was seven.

In 1969 the group was divided into two projects. One, headed by Johnston, was devoted to applied research concerning the control of subterranean termites and powderpost beetles. The other, headed by Smythe, was devoted to developing new, safe, and effective control techniques based on studies of the biology, behavior, and physiology of subterranean termites and powderpost beetles.

After Johnston retired in 1971, Smith served as project leader for the control part of the work until 1973 when the two projects were re-combined with Smythe serving as project leader. However, Smythe transferred to Washington in 1974 and Dr. Michael I. Haverty was appointed as project leader in 1975. Also in 1975, Dr. Ralph W. Howard joined the project as a chemist. Mauldin, the current project leader, was appointed in 1977 after Haverty transferred to California. Other personnel changes during the last 10 years include: Carter retired in 1984 and was replaced by Dr. C. A. McDaniel in 1985; Howard transferred to Manhattan, Kansas in 1984; Smith retired in 1980 and was replaced by Dr. Susan Jones in 1981; Beal retired in 1986 and was replaced by Dr. Bradford Kard in 1987. In January of 1990, a wood technologist, Ms. Maureen Mitchoff, joined the unit.

Results from research conducted by this project have been used to protect millions of wooden structures and wooden products from damage by subterranean termites and wood-destroying beetles. However, the scientists in the unit continue to search for new and safer chemicals and methods for protecting wood because of the controversy about the chemicals now in use. Based on data from the Gulfport unit, the Environmental Protection Agency has registered and approved labels for five chemicals during the 1980s. These chemicals are chlorpyrifos (Dursban[R]), cypermethrin (Demon[R]TC), fenvalerate (Tribute[R]), isofenphos (Pryfon[R] 6), and permethrin (Torpedo[R] and Dragnet[R]).

Currently, the unit is continuing to search for alternative soil treatment chemicals, evaluating borate compounds for wood protection, testing a bait-toxicant method of termite control, and extracting and identifying chemicals from naturally termite resistant woods for evaluation as wood protecting chemicals.

Imported Fire Ants

Clifford S. Lofgren

Introduction

Countless numbers of exotic insects have become established in the United States, but few have been as successful, nor had as much impact, as the red and black imported fire ants (IFA), *Solenopsis invicta* and *S. richteri*. They are characterized by their large mounds or nests and aggressive stinging behavior, which, has caught many unwary individuals by surprise and quickly revealed the reason for the "fire" in their common name. Their native habitat in South America is along some of the major river systems that flow south and north through the center of the continent. *S. invicta* occurs along the Paraguai River from Santa Fe, Argentina up to and beyond Caceres, Mato Grosso, Brazil. It has also been reported from Porto Velho, Rondonia, Brazil on a tributary of the Amazon River. *S. richteri* occurs to the north of Buenos Aires, Argentina and in Uruguay.

Over 1,000 scientific papers have been published on their biology and control. In this short paper I will limit myself to highlights of their (1) history in the United States; (2) biology and behavior; (3) economic impact, and (4) methods of control.

Historical Perspective

IFAs were transported on ship's cargo to the port of Mobile, Alabama, in the early 20th century. The first report of their presence is attributed to Loding (1929) who found them girdling the bark of young satsuma orange trees in orchards and nurseries near Mobile. At this time they were identified as *S. saevissima* var *richteri*. They increased in population rapidly and by the late 1930s they were so abundant in nearby Baldwin County that the first organized effort was made to control them on cropland. A survey made in 1949 revealed their presence in 28 counties: 12 in Alabama, 14 in Mississippi, and 2 in Florida. Infestations were found over 100 miles from Mobile at Meridian and Artesia, Mississippi, and Selma, Alabama. By 1953 they had been detected in 102 counties in 10 states. This major long distance spread resulted from transport of small colonies or queens harbored in soil attached to nursery stock and grass sod. Once established in an area, their spread through mating flights soon expanded the area of infestation.

During the late 1940s, black and red forms of IFAs were noticed in the Mobile area. Later, the 2 forms became geographically separated. The red form occupied over 90% of the infested territory while the black form occurred only in northeastern Mississippi and northwestern Alabama. In the early 1970s Buren (1972) concluded that they were separate species and gave them their current names. More recently, viable hybrids have been found in northern Mississippi, Alabama, and Georgia, suggesting that *S. invicta* and *S. richteri* may be a single species, Vander Meer *et al.* (1985).

The rapid spread of IFAs, and the appearance of large numbers of their huge mounds, alarmed farmers and urbanites alike. Public pressure for their control led to congressional implementation of a Federal-State program for control and eradication in 1957. Early control attempts with heptachlor and dieldrin effectively reduced IFAs populations, but damage to nontarget wildlife brought on strong environmental protests and their use was abandoned in 1963 following the development of mirex bait (Lofgren *et al.* 1963). This highly effective bait, composed of a toxicant (mirex) a food attractant (soybean oil) and a granular carrier (corn cob grits), was used in a series of control and eradication trials in the late 1960s and early 1970s. The preferred treatment regimen consisted of 3 applications of 1.25 lbs. of bait per acre (1.7 g. toxicant) over a period of 1 1/2 to 2 years (Banks *et al.* 1973). While this series of treatments was effective, the program was fraught with financial, organizational and environmental problems. It was finally discontinued in the mid 1970s when residues of the chemical mirex were evident in a wide range of nontarget animals, including man. The overall Federal-State program continued and concentrated on detection and elimination of new isolated infestations and quarantine enforcement (Lofgren 1986b).

Biology and Behavior

IFA abound in the tropics and subtropics of their homeland, but they appear even more suited to the warm wet climate of the southern U.S.A. *Solenopsis invicta* occupies over 95% of the infested acreage with population densities of over 100 colonies per acre in some areas. Their success is attributed to (1) a high reproductive rate, (2) effective foraging behavior, and (3) adaptability to a wide range of habitats. IFAs have been likened to weeds (Tschinkel 1986) in that they quickly invade disturbed habitats. New roads, building sites, and farmland are prime targets for establishment of colonies by newly mated queens.

The life cycle of a colony begins when the winged sexuals leave their mounds on mating flights, which usually occur in the early afternoon on warm days, 1 to 2 days following a rainfall. The males and females mate at altitudes of 300 to 500 ft. Flights are most frequent from May to July, but have been recorded in every month. After mating the male dies and the mated queen alights, chews off her wings and excavates a small nuptial chamber 2 to 3 in. deep in the soil. She lays her first eggs within 2 days. They hatch in about 7 days. The larvae are fed by the queen from food reserves stored in her body and pass through 4 developmental stages before they pupate. Developmental time from egg to adult is 3 to 4 wks. at temperatures of 80° to 90°F.

The first workers are very small and called minims. They assume the tasks of queen and brood care and foraging. The queen controls the behavior of these workers by release of attractants or pheromones. She eventually becomes an egg-laying "machine" capable of producing her own weight in eggs (2,000 to 3,000) every 24 hrs. Colony growth is rapid under the warm climate of the south and it is not unusual for a colony to contain ten to twenty thousand individuals in 4 to 5 months. However, individual queens and young colonies face many hazards from extremes of weather and predators. Probably less than 1 in 1,000 queens survive long enough to establish a mature colony (1 to 2 hundred thousand ants). This poor success rate is compensated for by the production of large numbers of sexuals. As many as one hundred thousand queens per acre may take flight in generally infested areas (30 to 50 mounds per acre).

The colony mound, or nest, is dome or cone shaped and honeycombed with tunnels that may extend 10 ft into the soil. Their size varies depending on soil type, landscape and rainfall. The mound serves for protection and thermal and moisture regulation. The workers continually move the queen and brood to achieve optimal conditions for growth (80° to 90°F and 70 to 90% R.H.).

Foraging tunnels radiate outward from the nest a few inches below the soil surface. They may often extend 50 to 70 ft. Foraging occurs through exit/entrance holes spaced at irregular intervals. This network of tunnels, and a complex foraging behavior controlled by physical and chemical attractants, allows maximum foraging efficiency (Vander Meer 1986). IFAs have a varied diet. They prey upon small animals, dead or alive, and quickly swarm over and sting to death slow moving organisms such as caterpillars and earthworms. Their high activity rate and energy consumption requires large amounts of carbohydrates which they obtain from plant sap, germinating seeds and honeydew produced by aphids, scales, and mealybugs.

Medical and Agricultural Impact

IFAs impact is directly related to their aggressive nature while foraging or defending their mound. The medical problems induced by their behavior are well known. Once a person is stung the reason for the word "fire" in their common name is easily understood. Dozens or hundreds of ants can quickly crawl onto a leg or arm placed in the wrong spot. Their venom consists of over 95% chemical alkaloids which kill the cells where they are injected causing a characteristic pustule. The venom also contains small amounts of 3 to 4 allergenic proteins to which some persons develop severe hyperallergic reactions. An estimated 67 to 85 thousand persons per year require the care of physicians and/or emergency treatment for anaphylactic shock (Paull 1984; Adams 1986). Deaths have occurred when appropriate treatment was not obtained.

Until recent years, IFAs damage to crops was considered insignificant. However, several instances of severe damage have been reported during this decade. These crops include citrus trees, soybeans, sorghum, potato tubers, eggplants, corn, okra, hay, pecans, and pine seedlings. Severe damage has occurred in young citrus groves in Florida where up to 25% or more of 1 to 4 year old trees in some groves have been girdled and killed when the ants fed on their plant sap. Greatest monetary loss in the south may occur to soybeans where losses of 5 to 6 bushels per acre have been documented (Adams *et al.* 1983). Indirect damage to pecans occurs when IFAs protect aphids and mealybug populations from predators while feeding on their honeydew secretions (Tedders 1989).

A variety of other problems are created when IFAs invade telephone and electrical junction boxes, meters, and air conditioners. The ants chew insulation from wires, short circuit switches, and fill the devices with dirt. Many reports of death or injury to wildlife and domestic animals have been recorded (Lofgren 1986a).

Because of the varied diet and aggressiveness of IFAs, it is not surprising that some of their activity is beneficial. Sugarcane growers in south Louisiana view them as an essential component of a predator complex on the sugarcane borer. Predation on other pest arthropods include ticks, earwigs, boll weevils, bollworms, horn flies, cucumber beetles, and tobacco budworms (Reagen 1986).

Control of IFA

Pest ants can be controlled most effectively with baits consisting of a slow acting chemical toxicant combined with a food attractant. For outdoor control, these components are generally combined with a granular carrier for ease of distribution. Effective control requires that the toxicant reach the reproductive queen. Unless she is killed or sterilized, the colony may survive even though most of the workers are dead. Mirex bait, which was mentioned previously, was highly effective for IFAs control. Since its demise, three new baits have been commercialized. Of the three, Amdro® gives the fastest colony elimination and is preferred for urban situations (yards, schools, parks). The toxicants in the other 2 baits, Logic®, and Affirm®, are classified as insect growth regulators. These compounds act by inhibiting queen reproduction and have little effect on workers present when the colony ingests the bait. Consequently, 2 to 4 months may be required before all workers die. These baits are more suitable for agricultural or non-urban applications where rapid colony elimination is not a prerequisite (Lofgren 1986a).

Control of IFAs can be obtained with direct application of chemicals to the mound. To be effective, these mound drenches, fumigants, or granular formulations must be applied in such a way as to directly contact as many ants in the mound as quickly as possible. Inherent problems are mound soil type, wetness and temperature, queen location in the mound, and rapid evacuation of the mound by surviving workers. All of these factors influence the possibility of contacting and killing the queen. Satellite colonies may appear and consequently need to be treated. This may occur also with bait applications.

Research on biocontrol of IFAs is in progress in South America. Several diseases and parasites of IFAs have been discovered, but to date, none appears promising for large scale reduction of IFAs populations (Jouvenaz 1986).

Future Considerations

Spread of IFAs to western states, particularly Arizona and California is a continuing concern. Several small infestations have already been detected and eliminated.

The discovery of polygynous or multiple queen colonies is a recent development that is of concern. In these colonies, numerous reproductive queens are tolerated and the workers of separate mounds are not aggressive to each other. Mound densities of several hundred per acre are common and worker numbers 2 to 4 times that associated with single queen colonies have been observed. It follows then, that impact on man, agricultural land, and wildlife is increased many times.

In conclusion IFAs will be with us for years to come and we must continue to develop safe and effective methods for their control in areas where they cause problems.

Suggested Further Reading

Adams, C.T., W.A. Banks, C.S. Lofgren, B.J. Smittle and D.P. Harlan. 1983. Impact of the red imported fire ant, *Solenopsis invicta*, on the growth and yield of soybeans. J. Econ. Entomol. 76:1129-1132.

Adams, C.T. 1986. Agricultural and medical impact of the imported fire ant. pp. 48-57. *In* C.S. Lofgren and R.K. Vander Meer (eds.) Fire ants and leafcutting ants: Biology and management. Westview Press, Boulder Colo.

Banks, W.A., B.M. Glancy, C.E. Stringer, D.P. Jounvenaz, C.S. Lofgren and D.E. Weidhaas. 1973. Imported fire ants: Eradication trials with mirex bait. J. Econ. Entomol. 66:785-789.

Buren, W.F. 1972. Revisionary studies on the taxonomy of the imported fire ants. J. Georgia Entomol. Soc. 7:1-26.

Jouvenaz, D.P. 1986. Diseases of fire ants: Problems and opportunities. pp. 327-338. In C.S. Lofgren and R.K. Vander Meer (eds). Fire ants and leafcutting ants: Biology and control. Westview Press, Boulder, Colo.

Loding, W.P. 1929. An ant (Solenopsis saevissima richteri (Forel). U.S. Dept. Agric. Insect Pest Surv. Bull. 9:241.

Lofgren, C.S., F.J. Bartlett, C.E. Stringer and W.A. Banks. 1963. Imported fire ant toxic bait studies: Further tests with graulated mirex-soybean oil bait. J. Econ. Entomol. 57:695-698.

Lofgren, C.S. 1986a. Economic importance and control of imported fire ants in the United States. pp. 227-256. S.B. Vinson (ed.). In Economic impact and control of social insects. Praegar Publishers, New York, N.Y.

Lofgren, C.S. 1986b. History of imported fire ants in the United States. pp.36-47. In C.S. Lofgren and R.K. Vander Meer (eds.). Fire ants and leafcutting ants: Biology and control Westview Press, Boulder, Colo.

Paull, B.R. 1984. Imported fire ant allergy: Perspectives on diagnosis and treatment. Postgrad. Med. 76:155-161.

Reagen, T.E. 1986. Beneficial aspects of the imported fire ant: A field ecology approach. pp 58-71. In C.S. Lofgren and R.K. Vander Meer (eds). Fire ants and leafcutting ants: Biology and control. Westview Press, Boulder, Colo.

Tedders, W.L., C.C. Rielly, B.W. Wood, R.K. Morrison and C.S. Lofgren. 1989. Impact of IFA on pecans in Georgia. pp. 64-78. In Proc. 1989 Imported Fire Ant Conference, Biloxi, Miss., April 18-19, 1989. Univ. of Georgia, Athens, Ga.

Tschinkel, W.R. 1986. The ecological nature of the fire ant: Some aspects of colony function and some unusual questions. pp. 72-87. C.S. Lofgren and R.K. Vander Meer (eds.). In Fire ants and leafcutting ants: Biology and control. Westview Press, Boulder, Colo.

Vander Meer, R.K. 1986. The trail pheromone complex of Solenopsis invicta and Solenopsis richteri. pp. 201-210. C.S.Lofgren and R.K. Vander Meer (eds.). In Fire ant and leafcutting ants: Biology and management. Westview Press, Boulder, Colo.

Vander Meer, R.K., C.S. Lofgren and F.M. Alverez. 1985. Biochemical evidence for hybridization in fire ants. Florida Entomol. 68: 501-506.

Cockroaches

Philip G. Koehler and Richard S. Patterson

Cockroaches evolved more than 250 million years ago and are considered a primitive but successful insect. In the Carboniferous period, cockroaches were one of the most abundant life forms in humid areas associated with ferns and palms. In fact, the situation today is quite similar, with cockroaches abundant in wet tropical areas of the world. Although considered primitive, cockroaches have had geological epochs to evolve efficient mechanisms for survival and reproduction. Their ability to survive is remarkable. For example, we placed 3,000 German cockroaches in a 20 gallon metal tub with food, water, and harborage. But the food we fed them contained no protein. Three years later, the cockroach colony was still alive, and females were still maturing and hatching egg capsules. Survival of a select few cockroaches was assured by cannibalism and a highly evolved method of recycling nitrogenous wastes for protein synthesis.

Of the more than 4,000 species of cockroaches in the world, only 66 are reported in the United States, and only about a dozen are considered pests. Although cockroaches are plentiful in the tropics, they are not usually a food source for primitive cultures because most species have foul smelling and bad tasting secretions. Even though they are not used for food, they are often accidentally consumed with food. Cockroaches have been used by man for various purposes. They are the "laboratory rat" for many students, entomologists, and physiologists. The oriental cockroach has been used as fish bait. Cockroaches have been advocated as cheap food for chickens, medicine for humans, and pets for the person who has everything. Sailors have used them for hundreds of years as the thoroughbreds of shipboard races.

Cockroaches are almost universally despised by people. Children up to the age of about four years have no aversion to cockroaches, but are quickly taught by parents that cockroaches are filthy and should not be touched or put into their mouths. Behavioral studies have shown that when plastic model cockroaches were placed into a child's drinking glass, children less than four readily drink the fluid; whereas children over four would not put the glass to their mouths. Such studies readily demonstrate that cockroach aversion is learned, not inherited.

The name "cockroach" has become almost universally synonymous with uncleanliness and disease. As a result, people with cockroach problems often call them by other names, such as "water bugs," "croton bugs," "Bombay canaries," or "palmetto bugs." I have displayed cockroaches at shopping malls in Florida where people were shocked to learn that the Florida palmetto bug was actually a cockroach. They would indicate that the Chamber of Commerce had assured them that the palmetto bug was not a cockroach, and on that assurance, had bought Florida real estate.

It is also interesting to observe that no nationality wants to be associated with cockroaches. The German cockroach is not from Germany, but was derogatorily associated with Germans. In fact in Germany, this species is called the Italian cockroach even though it is originally from Africa or Asia.

Cockroaches are considered pests because they consume and contaminate food, are capable of carrying disease, have an offensive odor, and can cause allergies. Cockroaches can destroy electronic equipment, such as computers, by causing short-circuits with their bodies or chewing through wiring and causing electrical fires. In heavy infestations, cockroaches will even try to harbor in people's ears at night or chew off calluses on hands and feet and gnaw on eyelashes. Children put to bed with milk or food smeared on their faces often wake up with red welts and clean faces as a result of foraging cockroach hordes.

It is interesting that in the laboratory, we pride ourselves for raising large numbers of cockroaches. But there are many people in the United States who do a superb job of raising cockroaches in their own homes. In a recent survey of low income housing in the southeastern U.S.A., half of the apartments had more than 15,000 German cockroaches per apartment, and 95% had intolerable levels of infestation. Heavy infestations could be detected at the door because of the cockroach odor. Inside the kitchens, cockroach excrement contaminated dishes, pots, pans, and eating utensils. In bathrooms, cockroaches were numerous around the toilets and bathtubs. In bedrooms, cockroaches were harboring near beds so that they could crawl onto sleeping individuals and feed around eyes and ears. One boy claimed that in order to sleep at night, he had to put cotton in his ears to keep the cockroaches out.

Cockroach research, nevertheless, is enjoyable and rewarding. There is no shortage of new and interesting discoveries. However, conducting field work can at times be dangerous. In 1983, we modified our sampling procedures for determining cockroach populations in apartments. Prior to 1983, we had always visually inspected drawers, cracks, crevices, and cabinets and counted the cockroaches. But residents in the 1980's became less cooperative, and we became more uncomfortable looking through all their belongings. Often we would find automatic weapons, illegal drugs, and other paraphernalia that the residents did not want us to find. Now we have gone to the placement of roach motels in apartments for 24 hours to determine the numbers of cockroaches. We spend as little time as possible in apartments and avoid going through any private belongings. Law enforcement agencies, we believe, have used us in the past to gain access into apartments. Occasionally, the janitor who opens the door for our technicians is an undercover policeman who is interested in obtaining information about illegal drug sales.

The presence of illegal drugs has also changed the types of conditions we see on a daily basis. Although the vast majority of residents are cooperative and want some method of control for what they feel is a serious problem, we have seen children less than one year old abandoned in apartments, child abuse, spouse abuse, and drug abuse. For these reasons there are few researchers who are willing to do field work with cockroaches.

Despite all these problems, we have been active in developing new technologies for cockroach control. We have been involved with the development of birth control for cockroaches (hydroprene and fenoxycarb), bait trays for odorless and hazardless cockroach control (hydramethylnon and sulfluramid baits), pyrethroid insecticide formulations for the pest control industry (cypermethrin, esfenvalerate, and cyfluthrin), and mass production of cockroach egg parasites as biological control agents for inundative releases. These new technologies have resulted in superior control products that are now available as over-the-counter pesticides or new technologies for the pest control industry. In 1989, cockroach problems were dramatically reduced due to the introduction of these new technologies. In fact, it was almost impossible to find heavily infested apartments throughout the country for research programs.

These new technologies do not spell the end of the cockroach. Its ability to exist over the past 250 million years assures us that the cockroach will evolve mechanisms or behaviors to survive these new technologies. For every advance we make against the cockroach, it makes one to survive. We need to remain one step ahead to maintain the ability to control cockroach problems and reduce the indignity and health risks of explosive infestations of cockroaches.

Suggested Further Reading

Bell, W. J. & K. G. Adiyodi. 1981. The American cockroach. Chapman and Hall, London.

Bennett, G. W. & J. M. Owens. 1986. Advances in urban pest management. Van Nostrand Reinhold, New York, NY.

Cornwell, P. B. 1968. The cockroach. A laboratory and industrial pest. Vol. 1. Hutchinson, London.

Cornwell, P. B. 1976. The cockroach. Insecticides and cockroach control. Vol. 2. Associated Business Programmes, London.

Guthrie, D. M. & A. R. Tindall. 1968. The biology of the cockroach. William Clowes and Sons, London.

Africanized Honey Bees in North America[1]

Roger A. Morse

In 1956, Professor Warwick E. Kerr of Brazil brought 132 queen honey bees from Africa into his country. He did so at the request of the Brazilian Ministry of Agriculture, which was encouraged in this direction by Brazilian beekeepers. In that year Kerr had won the "Andre Dreyfus National Prize in Genetics" and used the money to study in Africa. Beekeepers in Brazil were aware that there was almost no practical beekeeping in the warmer tropical regions of their country. The climate there is too warm for the European honey bees that had been introduced into the temperate and subtropical parts of the country more than a century earlier. It was known too that honey bees in Africa were good honey producers, and though aggressive bees, were used by commercial beekeepers in many parts of that continent. The introduction has been a great success in Brazil and in many areas where there was once no beekeeping there is now a flourishing industry.

Prior to the introduction of the African bees little honey was produced in Brazil. Figures published by the Food and Agriculture Organization of the United Nations show a steady increase in honey production in Brazil during the past several decades. Beekeepers who have visited the country are impressed with the size and management they have observed on the part of many beekeepers who operate between one and four thousand colonies. Honey production in 1988 was reported to be in the vicinity of 35 million pounds.

Origin of the word "Africanized"
We call the new bees in Brazil "Africanized" because they have interbred to some extent with the local European honey bees. The honey bees we know in the U.S.A. are natives of Europe, Africa, and the Near East and were carried to other places, including North and South America, by the early European settlers who immigrated in the 1600s. The honey bees now in the north of Brazil, the warmer part, are almost pure African while in the south where the climate is more temperate, there has been more mixing with European stock. The

[1]Reprinted with permission from *Bees and Beekeeping* by Roger A. Morse, Cornell University Press, Ithaca, New York, 1991.

bees in the north that are almost pure African are far more aggressive and do a much better job of defending their nests than do the mixtures further south though climate plays a large role in the difference.

Source of the African stock

There is no mystery about where the original queen honey bees taken to Brazil came from. Most of them were grown in an apiary a few miles south of the South African capitol city of Pretoria. The greatest number (120) came from a beekeeper by the name of W. E. Crisp, a small number from an E. A. Schnetler, both from South Africa, and one from Dr. F. Smith of Tabora, Tanzania. This is documented in a paper by Kerr (1967). Most of the colonies in the vicinity of the apiary that was the source of most of the queens near Pretoria are owned and operated today by a commercial beekeeper who has about 2,000 colonies and is a successful honey producer. One would not think this was possible if he or she believed the stories that we hear about these bees from South America. It is a great curiosity that the scare stories that have accompanied the spread of the Africanized honey bees north toward the U.S.A. are not heard in Africa or Brazil today. Not all of the queens taken into Brazil survived the journey and their introduction into colonies in Brazil. As a result the number of queens given as being imported is different in various papers.

The history of beekeeping in Africa

Beekeeping in many parts of Africa is an art that predates modern history. We are aware that the ancient Egyptians sailed up and down the east coast of Africa trading in slaves, gold, frankincense, myrrh, honey, beeswax, and other products, as many as 5,000 and perhaps more years ago. The methods of beekeeping in some of the eastern African countries, which appear to be some of the best honey producing areas in Africa, are not so sophisticated as in some parts of the world; however, in some years the countries of Ethiopia, Kenya, Somalia, Tanzania, and Uganda will collectively export as much beeswax as we produce in the U.S.A. South Africa has a flourishing beekeeping industry, and other parts of Africa, all using these so-called ferocious bees, are also gaining expertise in beekeeping.

Effects of the introduction into Brazil

When the Africanized honey bees were first introduced into Brazil in the state of São Paulo, local beekeepers were accustomed to using small pipe smokers and sometimes wore no veils nor took any special precautions in handling their European bees. These beekeepers manipulated their colonies in much the same way as do many European and American beekeepers. With the Africanized honey bees, they soon found they needed to dress more carefully and use smokers that delivered a greater volume of smoke to calm the bees. They found too that it was advisable to keep their colonies in apiaries remote from people and domestic animals. However, even beekeepers in the U.S.A. take precautions to keep their bees in locations where they will not be a nuisance. Manipulating bees or harvesting honey from a large commercial apiary in the U.S.A. can cause the bees to become greatly aroused and to sting anyone or any animal in the vicinity.

Spread of the Africanized honey bees north and south

Professor Kerr was soon aware that the bees he had imported were different and were bringing about changes in beekeeping in Brazil. In 1969 he published a map showing how the bees had spread both north and south during the 11 years from 1957 to 1968 from the place where they had been introduced. Kerr's map shows the bees spread north, into warmer areas, more than three times as rapidly as they spread south. Several years later, using Kerr's data as a starting point, Taylor (1985) projected in 1975 that the bees would reach the U.S. between 1988 and 1994 though natural migration; his prediction has proven to be remarkably accurate.

Africanized honey bee biology

The biology of the Africanized honey bees at the front of those migrating north has been little studied. In most ways they are much the same as European honey bees; however, there are some interesting differences. My students discovered in 1988 that these nearly pure African honey bees have a special dance that they use to signal the other bees in the swarm that they should become airborne and move a long distance. Our research was done near the southwestern Mexican city of Tapachula near the Guatemalan border as the Africanized bees were moving through that city. We placed an ad in the local newspaper offering a reward to anyone who would phone and tell us about a migrating swarm that was settled near their home. We received a number of phone calls and investigated and worked with many swarms within the city limits.

The migrating swarms appeared to settle in a tree or bush, in the shade, in mid-afternoon. At that time scout bees became airborne and searched the vicinity for nectar. As food sources were identified these scouts returned to the clustered swarm and danced to recruit other bees to seek and collect the food they had found. This foraging continuing until dusk and resumed early the next morning as soon as it was light and warm enough for the bees to fly. Successful foragers gave the food they had collected to others in the swarm that acted as storage tanks for the collected honey. We were able to observe the dances performed by the foodseeking scouts and found they were not different from those performed by scouts from swarms of European honey bees with which we had much more experience. In early mid-morning, when the bees had tanked-up sufficiently, the scouts started a new and much slower dance than we had ever seen before. After several minutes, when all of the scouts were in agreement and dancing the same dance, the swarm became airborne and moved in a northwestern direction. It was now in a migratory mode and with its food reserves was capable of moving many miles. It has been in this manner that Africanized honey bees, though successive swarms, have been able to migrate 200 or more miles in a single year and have spread themselves over South America so rapidly. We do not have data to indicate how far an individual swarm of Africanized honey bees might fly, however, observations on African honey bee swarms in Kenya shows they may move many miles in a single day.

Brazil's military government and the role of politics

To return to early phases of this African introduction, it is important to realize that on April 1, 1964, military forces took over the government in Brazil. Professor Kerr, who had introduced the African bees was a well-known scientist in Brazil and had represented his country at many international meetings; in this regard he was badly needed by the government. However, Kerr was also critical of the military government and there was conflict between him and the local military commander. Kerr was jailed twice by the military, the first time in 1964 when he protested that a group of local railway workers were being maltreated, and a second time in 1969 for protesting the torture of a Catholic nun. In an effort to discredit Kerr as a scientist, the local military played upon the fear that many people have of stinging insects. Since most people do not know the difference between bees and wasps, any stinging incident, many of which were caused by wasps, was blamed on Professor Kerr.

The Brazilian military called the bees, in Portuguese, the language of Brazil, *abelhas assassinas* (killer bees). So far as I can determine, the first mention of the words "killer bees" in the U.S.A. was in Time Magazine in the September 24, 1965 issue that picked up one of these military press releases. Much the same story was repeated in a second article in the same magazine in the April 12, 1968 issue. Those stories prompted others to write in this same vein and the term, and the Brazilian association with "killer bees", became firmly established and continues to live. Several horror-type movies with titles such as *The Savage Bees, Terror out of the Sky, The Swarm,* and *The Killer Bees* have been produced by Hollywood. There have been a number of similar articles about these bees by popular writers, with no experience in beekeeping, seeking to capitalize on the theme. I have no doubt there will be more such movies and papers.

Predicting the future

In order to determine what will happen as the Africanized honey bees spread in North America, it is well to look at the situation in Brazil, which now has returned to civilian rule, and at countries such as Argentina and Peru where the Africanized bees have become firmly established. Argentina has long been one of the major honey exporting countries in the world. The vast flat, rich land of the Pampas that occupies much of central Argentina is covered with clovers and thistles that produce a rich supply of nectar that the bees make into some of the finest table honey known. Honey bottlers and distributors from around the world are very familiar with the high quality of Argentina's honey. It is much sought after and receives a premium price on the world market. The Pampas lie in a temperate, not tropical area and there the European honey bees, not the Africans, thrive. Many Argentinean beekeepers migrate north, just as many of our U.S.A. beekeepers migrate south, to grow young queens and bees in the spring that will be later be used for honey production. When these Argentine beekeepers go north they enter warm areas where Africanized honey bees grow and thrive; obviously African genes are mixed with those of the European bees when they are again carried south. Still, the African bees have not had an adverse effect on the Argentinean beekeeping industry and we hear no "scare" stories from that country.

Proof that Africanized honey bees have not had an adverse effect on honey production in Argentina is found by examining the Food and Agriculture Organization honey production figures for that country. Kerr *et al.* (1982) reported that Africanized swarms of honey bees were found in northern Argentina in 1968 and soon thereafter in the province of La Pampa, which is the leading honey producing area in the country. During the period 1960 through 1979 there was approximately 49 million pounds of honey produced in Argentina annually; the range during this 29 year period was 31 to 66 million pounds. There was a sudden increase in production in the 1980's. In the nine years from 1980 through 1988 production averaged 75 million pounds; the range was from 55 to 88 million pounds. There is no question that honey production in the 1980's was stimulated by a strong demand and higher prices worldwide; however, the chief point is that production was not adversely affected because of Africanization of the bees.

A paper on beekeeping in Peru records events in that country as it was invaded by African bees that replaced the European honey bees at the lower elevations (Kent, 1989). Peru is the home of the northern Andes mountains; the African bees have not been found at the higher elevations. Peruvian beekeepers did not count African honey bees among their most serious problems when asked to list what they thought might be done to help and encourage the industry. The number of stinging incidents in Peru was not significantly different from the times before the Africanized honey bees entered the country.

Effects of Africanized honey bees in the U.S.A.

Beekeepers in the U.S.A. will soon learn how to manage these immigrant bees. Some will give colonies of Africanized bees new queens so as to change their genetics. Some beekeepers will abandon their hobby or vocation because they will find that working Africanized bees means they will be stung more often. Beekeepers in the southern states, where the bees will be the most common, will learn to dress more carefully to protect themselves. They will wear boots, not shoes. They will use suits with zippers, not buttons, to protect against bees that will otherwise crawl between the buttons. They will select their apiary sites with greater care. For the most part the general public will not be aware of any change.

There is no doubt that officials in many towns, villages, and cities will think about enacting ordinances banning keeping honey bees within their boundaries. This has always been a small problem for beekeepers, and we are only now beginning to collect data on the number of honey bees and wasps colonies that live in buildings, hollow trees, and caves. We are finding there are far more colonies of honey bees living under these wild conditions than we had realized. These insects are where they are because food is available for them. The only way one can effectively reduce the number of colonies of stinging wasps and bees in an urban area is to eliminate their food. Clearly, almost no one will want to prohibit the growing of flowering plants on their property yet this is the only way to discourage these insects from living where they do. The next question is, is it better to have colonies in man-kept hives, under the control of beekeepers in an urban area, or is it best to leave the foraging to wild, unattended colonies? No doubt there will be different viewpoints on this question. Many of these feral colonies nest high in the air in trees and buildings so that their flight lanes are above the heads of those walking in the vicinity, thus the colonies themselves are usually no problem. However, the data we have indicates that feral colonies live in much smaller cavities and produce more swarms, usually far more than twice as many, than do those colonies in man-kept hives.

After several years of argument and discussion, it now seems clear that researchers, extensionists, and beekeepers are finally aware that nothing has been done, or can be done, to stop or slow the advance of the Africanized honey bees into the U.S.A. Most people are now realizing that their coming is inevitable; however, this was not until a vast sum of money was spent in Mexico in an effort to stop the northward migration of the Africanized honey bees. The question is now how should we react and how can we gain from what is happening?

Do Africanized honey bees have virtues?
We are already aware that African and Africanized honey bees have some virtues. They are resistant to several diseases that plague beekeepers in other parts of the world. The dreaded varroa mites, Asian in origin, devastate colonies of European honey bees wherever they come into contact with them. They are found in every colony in Brazil where we look for them. However, Brazilian beekeepers do not treat for varroa disease because mite populations never become very high; apparently the Africanized honey bees have a certain degree of resistance to the mites. Perhaps because of better grooming practices. Beekeepers in Africa and, the warmer parts of South America, do not complain about the several common honey bee diseases, though American foulbrood has been found in Argentina recently and may prove to be a problem there. In the subtropical parts of Brazil we know diseases were a problem when only European honey bees were present.

African and Africanized honey bees build populations rapidly, in fact, more so than do European honey bees. The queens in African and Africanized honey bee colonies lay more eggs in a given period of time compared to European bees bee colonies and the time required for the development of worker bees is slightly shorter. This is usually a plus but it also means beekeepers must be alert to their colonies needs and give them room for expansion; if this is not done on time, the Africanized bees will swarm excessively. However, with proper management beekeepers may capitalize on this rapid colony growth and use it to their advantage to produce more honey. There is great variation within races of Africanized honey bees regarding their tendency to swarm and careful selection of stock is required when requeening colonies.

The role of climate on Africanized honey bee behavior

We are aware that weather and climate have a profound effect on honey bee behavior, both in European and African bees. In the 1950s, a British researcher working in the East African country of Tanganyika (now Tanzania) recorded that colonies under tin roofs and "in other hot nesting places" were especially difficult to manage. However, colonies nesting in deep walls and cool places were placid. When bees were kept in specially constructed, well- ventilated bee houses it was possible to open and examine African colonies in much the same way as one would European colonies.

An experiment sponsored by the U.S. Department of Agriculture, and carried out by Brazilian researchers in Brazil, showed that when colonies were moved from a warm climate to a cooler one that the number of aggressive bees were fewer under cooler conditions.

The behavior of all races of honey bees is affected too by the quantity of nectar available to the bees. During times of dearth honey bees are much more defensive of their nests, presumably because of a greater danger of their being robbed by other honey bees. During a rain, when bees cannot fly and forage, they are also more defensive. A free-flying swarm that has exhausted its food reserve is called a "dry swarm" and bees in such swarms are inclined to attack any animal that comes near their nest. In describing or discussing aggressiveness in honey bees it is necessary to know and understand the effects these differences can have.

Summary

It will probably take five to ten years for beekeepers in the southern and warmer parts of the U.S.A. to learn how to best manage the Africanized honey bees. Unfortunately, as these bees have moved north in South and Central America only a few North American researchers have studied their biology. Researchers from this country have not studied how management will change and what steps should be taken to be able to gain the most from these bees. However, much of this knowledge is in the hands of Brazilian beekeepers, researchers, and extensionists who are increasingly being called upon by those in Central America for advice. Many commercial beekeepers from the U.S. have taken the time and trouble in the past few years to visit Brazil and talk with Brazilian beekeepers, and to work with their bees, so as to better prepare themselves for the coming of these bees. The Mexican government has put an extensive extension program in place in the Yucatan peninsula, which is a major honey producing area in that country. Increasingly, the American beekeeping industry is facing the fact that Africanized bees are migrating north and that a positive approach is necessary to cope with the new bees, especially the public's concern about them. We will continue to need and use honey bees to produce honey and pollinate crops. In time, the term "killer bee" will disappear and we will live with this much maligned creature.

Chapter 5

Controlling Insects of Agricultural Crops

THE FAR SIDE   By GARY LARSON

The Development of Insect Pest Management Chemicals

Larry L. Larson

Introduction

Since their beginnings, plants and animals have evolved the ability to fend off arthropod threats to their existence through a variety of complex chemical and biological approaches. It has only been within the last 50 years, however, that man has learned to augment nature with chemicals designed to protect our food, fiber, and the homes in which we live from the devastations of attack by various pests. Since World War II we have seen a trend toward more specific, more highly active pest control agents, many of which were patterned after already existing plant (pyrethrum) or microbial (abamectin) toxins.

Today, the pesticide industry is being pressured to replace the old broad spectrum insecticides with more specific, environmentally compatible agents with minimum crop residue. In general, the industry has responded with better studied, highly active, less toxic alternatives. In fact, modern pesticides are better studied and more understood than many drugs currently on the market, and rightly so! The United States enjoys one of the most productive agricultural system on the face of the earth. We must protect that system from any adulterant that may jeopardize the safety of our food supply.

The process of pesticide development is designed to detect key properties about a chemical early, while ensuring that improvements over current technology are made available around the world as expeditiously as possible.

This process is becoming more and more expensive (30-50 million dollars per commercial product) and because of this it is rapidly becoming the concern of large multinational corporations. This article will explore the typical development of pesticides within a global corporation. The hope is that it will let the reader better understand the complex process of research and development involved in the commercialization of a new pesticidal product and provide some insight into the motivation of the professionals involved in the process.

Early Research

The ideas for new pesticides come from creative biologists, microbiologists, biochemists, and chemists involved in early stage research. Two very different philosophies of discovery research are prevalent within the industry today. The first, non-design screening, has been responsible for many breakthroughs in pesticide science. This is the process where compounds not specifically designed as pesticides are screened against pests of agricultural importance. The source of these materials may include pharmaceuticals, chemical processes, fermentation products, etc.

The process of product discovery is exciting and full of serendipity. Most areas of novel chemistry have been discovered by chance. Juvenile hormones resulted from difficulties in rearing bugs on balsam paper towels. The benzoylphenylurea insect growth regulators were designed originally as herbicides.

Materials active in the screening process may serve as candidates for further optimization through the second type of discovery, process-design synthesis. Here specific biological targets are determined and given priority. Compounds are then designed and selected to produce the highest possible efficacy.

The primary screening is the beginning of product development. It is here that compounds are first evaluated and their spectrum of activity determined against key pest species. Very small quantities of compounds (a few hundred milligrams) are generally available at this stage so efforts are made to use them as efficiently as possible. In recent years, automation has taken over much of the drudgery of cranking through thousands of samples per year and the discovery professional has become much more sophisticated in looking at structure-activity relationships. The joy of scientific discovery is the key motivation for professionals involved in this arena.

Once a compound shows activity in the primary screening, they become candidates for secondary screening. In this process, the high rates used in the primary screening are titrated down to commercial levels, species spectrum is further evaluated, and symptomology noted. Some of these screenings may involve small outdoor plots to assess the impact of environmental parameters. These tests confirm that the compound still controls the pest under more natural conditions.

Before extensive field trials are undertaken, the physical properties and acute mammalian, aquatic, and avian toxicities are evaluated to assess the registrability of the compound. These tests determine the potential hazard if the chemical is accidentally swallowed, spilled on the skin, splashed in the eyes, inhaled, or spilled in the environment. On many occasions, a promising area of chemistry has been dropped due to toxicological concerns at this stage. Also, at this stage, a preliminary patent disclosure is filed in appropriate countries for the compound and/or its use.

Decision to Commercialize

Following field confirmation of the laboratory efficacy and preliminary toxicology, management must decide whether to commit millions of dollars and 7-10 years of additional research to register the product for use. Given the high resource commitment that comes with a decision to commercialize, a tremendous amount of information is required from company research scientists. Included in this analysis are answers to questions on product performance, market fit, potential market size, existing and experimental competitive products, patent position, safety of use, and registration timetable. Very few compounds make it to this decision point. It is estimated that over 10,000 compounds are screened for every product that makes it to this point. Due to escalating costs of product registration, larger and larger market opportunities are needed to justify commercialization. There are very few new market opportunities emerging, and the established large markets are becoming more and more competitive. This has led to fewer new candidates for commercialization over the last decade, and resulted in serious concerns over the future of tools for minor crop use. This situation has caused a resurgence of approaches which are less expensive to register and more environmentally acceptable to the general public. More consistent creative solutions are needed to solve these very serious concerns in the near future.

Predevelopment

Following the difficult decision to commercialize a potentially new product, the pesticide manufacturer begins a complex series of scientific studies using good laboratory practices, designed by the Environmental Protection Agency, to assure the safety and efficacy of the product in its potential labelled use. This involves extensive field testing to determine the spectrum and commercial fit of the chemistry. In this stage, label rates are determined. If the product involves food crop use, long-term animal toxicology studies are undertaken looking for effects such as cancer, birth defects, behavioral modification or reproductive effects. Under controlled conditions, animals are fed the material at very high doses to determine the highest level of the test substance that will produce no effect on laboratory animals. The environmental fate of the material is thoroughly studied under field use rates and conditions over 1 to 3 years, depending on the persistence of the material. Especially critical in these studies is a look at potential leaching of the material or its metabolites on a variety of soil types. In addition, studies under potential use conditions are instituted throughout the crop growing areas of the U.S.A. to determine efficacy and the potential levels of crop residues resulting from application of the pesticide according to label instructions. Safety of these levels of residue is established by determining the "No Observed Effect Level" (NOEL), the level at which the chemical has no harmful effect on the most sensitive test species in animal toxicology studies. The "Acceptable Daily Intake" (ADI) is then set by dividing the NOEL by an appropriate safety factor (up to 100 or more) depending on the type of toxicity. Thus, the ADI represents the level at which residues can be safely ingested by an average person over a lifetime without ill effects. The use of a safety factor means that the legal residue is far below that causing an effect in the most sensitive species.

"Experimental Use Permits" (EUP'S) may be issued by the EPA to allow more than 10 acres to be treated with the experimental product. EUP'S may be of two types depending on the existence of adequate residue data. If residue data are available a temporary tolerance can be applied for, leading to a crop non-destruct EUP. Otherwise the EUP will have a mandatory crop-destruct provision.

Development

Formal development begins as the last critical registration information is compiled and submitted to the EPA. While the agency reviews the data, universities and private cooperators test the experimental product under a wide range of field conditions in order to better understand the value of the product in the market. Large scale trials are undertaken with key dealers and distributors to prepare for the launch of the material in the market following its registration. The professionals involved in this work are motivated by a desire to solve customer problems, and reduce to practice the dreams of the discovery scientist. The development phase is an exciting time in a project, as the public at large becomes aware of a new tool and its advantages become more apparent in a pre-commercial setting.

The Registration Process

The "Federal Insecticide, Fungicide and Rodenticide Act" (FIFRA) and its amendments mandate the federal registration process for pesticidal products with agricultural or industrial uses. With congressional oversight, the EPA has promulgated 415 pages of implementing regulations designed to guarantee that no pesticide will be registered unless it performs its intended function without unreasonable adverse effects on the public or the environment.

In recent years, the EPA has been increasingly under attack by public interest groups due to perceived slowness to act in the public good. These groups have sometimes achieved the banning of pesticides on the basis of public opinion. The agency has courageously refused to take action until all the scientific facts are in. The EPA deserves the support of industry, and the general public, in the very difficult job they have of assuring the safety and abundance of our food supply.

In general, the EPA requires registrants to provide detailed information in the following areas:

Product Chemistry
Environmental Chemistry
Residue Chemistry
Hazards to Humans and Domestic Animals
Reentry Protection
Hazards to Wildlife and Aquatic Organisms
Hazards to Non-target Insects
Spray Drift Evaluation
Phytotoxicity to Target and Non-target Plants

The product chemistry section must identify impurities down to 0.1% concentration and establish limits for each one. In addition, a process description including starting materials, process conditions and some 20 physical and chemical properties of the active ingredients including solubility, color, odor, vapor pressure, and other factors is required.

The environmental chemistry section includes both laboratory and field studies. In the laboratory, degradation of the material is studied in air, soil, and water under the influence of light. Soil metabolism of the material is studied under anaerobic (oxygen depleted) and aerobic conditions. Finally, mobility or leaching in soil, and volatility from soil and plant surfaces are determined. In the field, dissipation and leaching of the material in the soil is studied under actual use conditions in 2-4 locations across the geographical areas of potential use. Accumulation of the material is studied in fish and rotational crops.

Residue data are required where the use pattern involves treatment of food or feed crops. The residue data must come from around 75% of the area of use on a state by state basis, and be balanced geographically to assure a proper overall spectrum of potential residues that may be detected under commercial use. Animal residue data are needed where detectable residues may find their way into animal diets. Process fractions are needed to determine if pesticide residues will concentrate in commodities such as refined oils from oil seed crops used to make vegetable oils. Metabolism studies are needed in major crops as well as animals to determine the ultimate fate of the pesticide in these organisms. These results quite often complicate residue studies due to the need to analyze for residues of significant metabolites in addition to the parent compound.

The hazards of the material to mammals must be determined by acute oral, dermal, inhalation, eye irritation, and skin sensitization tests. The chronic toxicity of the material must be determined in two rodent species over their lifetime with doses set to determine the maximum tolerated dose (MTD) and have at least one dose level without a measurable effect (NOEL). In addition, several mutagenicity tests are required as well as reproductive studies and teratology (birth defects) studies in two species. Using the results of these studies, and a study on the pattern of actual field use exposure, reentry, and protective clothing requirements are determined. This is an extremely important consideration for labor intensive crops such as fruits and vegetables, but any crop in today's intensive agricultural systems may be scouted, so reentry intervals are becoming of greater concern.

There are serious, larger concerns today with pesticides as potential hazards to wildlife and aquatic organisms. All prospective pesticides are required to pass an exhaustive series of tests to assure their environmental safety. Acute toxicology is required on mallard duck, bobwhite quail, bluegill (warm water), rainbow trout (cold water), honey bees, and water fleas (*Daphnia* sp., an invertebrate). The results of these studies, and the environmental persistence of the material, determine the need for further laboratory and field work. Some of the field studies that may be required can take up to 3 years and cost several million dollars.

Depending on activity profile, extensive studies may be required to evaluate spray drift and phytotoxicity to target and non-target plants. Recent experience with registration packages indicates that for a given material about 70 reports involving as many as 100 research people over 7 years or more are necessary to complete the submission. The final submission may require stacks of paper several feet high. It can take the EPA 1-2 years or more to evaluate the package and approve final labelling. Usually further studies are triggered which can take several more years to accomplish. In order to set a residue tolerance under the Federal Food Drug and Cosmetic Act, the EPA must be convinced by the applicant that the proposed tolerance level will be safe and measurable in raw agricultural commodities.

When the application is approved the EPA product manager prepares a final rule on the registration and the tolerance petition(s) for publication in the Federal Register. After a 30 day comment period, if no adverse comments are received, the residue tolerance becomes final and the label officially issues. At this point, very few years remain of the 17 year patent period to allow time to recoup some of the 30-50 million dollar investment before generic imitator products become legal.

Label Expansion

The labelling of a new insecticide marks the end of the first phase of the research process. It also marks the beginning of the label expansion phase. The search for additional uses for the material now begins, each of which must be reviewed by the EPA. Materials may change their character entirely as they go through this process. For example, chlorpyrifos insecticide began life as a specialty product for chinchbug control in lawns in the U.S.A., and tick control in cattle in Australia. In 20 years the label has been expanded to include dozens of food and fiber crops as well as structural pest control. Methods of application now include ground, aerial, and overhead irrigation.

The process of label expansion is not always an easy one. However, it is a very important process for the future efficiency of agriculture. Innovation with these chemical tools quite often is the result of practical experience. A good example is overhead sprinkler application of LORSBAN 4E on corn for European corn borer control in the Western corn belt. The original work in Nebraska demonstrated excellent improvements in activity with the technique due to more uniform deposition of the active ingredient than from ground or aerial application. However, much work had to be done to assure the technique would not pollute groundwater and that the residues did not exceed those previously established with already labelled applications. In addition, aerial applicators had to be convinced that we were sensitive to the potential effects on their business and would try to balance our approach. When all these potential problems were solved, the technique was added to the label and within a few years it has become the standard method of application in many areas of the corn belt at up to half the cost of previous treatments.

Product expansion professionals are motivated by seeing innovation in agriculture adopted before their eyes. Customer thanks become a great motivator for these specialists.

Reregistration

Reregistration is a process developed by the EPA to systematically review and update the data base for pesticides registered prior to 1978. This is necessary due to changes in requirements that have occurred over the years. This procedure has led to the withdrawal of many minor use products where the market size will not support the necessary studies. It also has overburdened the agency with hundreds of standards to be issued and mounds of reregistration data to evaluate over the next decade.

Conclusions

The process of pesticide research and development is critical to the continuation of our efficient system of food and fiber production in the United States. Many professional entomologists are employed in all aspects of this process. Their efforts will continue to offer creative solutions to many of the challenges currently facing pest control around the world. The process will ensure that no pesticide will be registered unless it performs its intended function without unreasonably adverse effects on man or the environment. Insect problems will continue to challenge entomologists far into the next century. We plan on being up to the challenge.

On the Outside Looking In on Entomology, Well, Almost —

William L. Hollis

For some, Fate offers very circuitous and mysterious routes into the future. Imagine, after five years of World War II and struggling to stay alive during combat tours in Europe or the South Pacific, finding yourself in a class room where Collembola, Cursoria, Corrodentia, Coleoptera, and twenty four other tongue twisting orders of insects, and their sub-orders too, had to be learned, pronounced, spelled, and understood. Well, my classmates and I shook our heads and asked again that big question: how did we get here?

The answer in my case was simple; it wasn't Fate this time. At least I didn't think it was fate when my new spouse informed me that I had a fine mind; it was brand new and had really never been used — Go Back To School. Before I could muster any opposition as the man of the house, I was further informed that I was essentially already registered and just had to report in. So I dutifully set off to discover the excitement, challenges, and gratification of working in the agricultural sciences. Whether it was fate or my wife doesn't matter, it was the best decision I ever made.

Going back to a University College of Agriculture didn't acquaint me with insects. I'd been combating them on the farm and then in Pacific Island jungles and was very familiar with them. What I learned in College was that they are associated with and the subject of a whole discipline, entomology, a new word I quickly learned. Entomologists are the practitioners of the discipline. They study insects, not bugs, and I accepted that.

We learned a great deal about entomology in those undergraduate classes and gained respect and appreciation for this science, and those who practice it. Some of my classmates were sufficiently intrigued to go on to become entomologists. In my case it was the plant sciences, physiology, and genetics, that drew me in another direction when we came to the crossroads of deciding on our professional future from among the many sciences and disciplines that comprise agriculture. So it was on to horticulture, or olericulture specifically, and food processing. It is from this vantage point and my experiences in food production and processing that I am offering some observations on entomology "From The Outside Looking In, Well almost." While entomology is no stranger to me, I'm not a professional entomologist.

It is interesting that some things one learns without knowing it specifically, or perhaps fully realizing it, until circumstances make it apparent. It was in this manner that I became aware, early on in my plant research, of how the success of any agricultural enterprise occurs only through the interdependence that exists, must exist, among the sciences and disciplines that attend agriculture, even those subjects that may seem remote. No one segment of agricultural science stands alone in practice. They all come together in the field and together help assure agricultural success. I soon found out that my plant science research was dead in the water and as much as a year of work could be lost without the help and advice of entomologists.

And so it is with farmers, the practitioners of plant science. They can't farm successfully without integrating entomology into their crop management practices. Although entomology is foremost in our minds, we must interrupt our story for just a moment to also recognize the integrated contributions of pathology, nematology, genetics, and the most recent addition to the state of the art — biotechnology (made possible by engineers), soil chemistry and physics, botany, plant physiology — and biochemistry, biophysics, nutrition, taxonomy, biometrics, engineering and other sciences that I'm sure my colleagues will tell me I left out.

All of these agricultural sciences have their valued stories to tell but the best I can do here is to nudge the non-agriculturally oriented folks who may read this book toward realizing that agriculture, as a biological industry, is the dynamic culmination of the shoulder to shoulder efforts of these sciences. When not done this way and when one or more workers in these disciplines don't look beyond their sciences, we have problems. Since none of us is perfect, I may find a few such instances to mention, as I look in on entomology.

I can sense what intrigues the entomologists when one realizes that the creatures being studied have, in many forms, flown, burrowed, crawled through incalculable generations from prehistoric times to be present now, and capture our attention. I can share, more realistically than vicariously, the challenges of understanding insects and the control of those we consider pests. It is obvious that insects are not going to be deterred by any means to stop their rapid evolution. They have the naturally endowed means of adjusting in succeeding generations to any control mechanism, natural or synthetic, by a process called resistance. The host, say a caterpillar of a parasite, is able to encapsulate the egg the parasite lays in its body and thereby render it harmless. The same is true of those parasites (hyperparasites) of the parasite. Also toxic bacteria and viruses or those that cause insect diseases, are eventually overcome in succeeding generations. Likewise, the synthetic organic pesticides and plant varieties are genetically altered to produce chemical metabolites toxic to an insect, the so-called insect resistance plant varieties.

Considering these natural circumstances, it is obvious why entomologists settled on the strategy of using integrated pest management (IPM). As I see it, the objective is to reduce the incident opportunity for insects to develop resistance against any one control measure and subsequently against more than one. Insects have the astounding ability to develop cross resistance. While it seems numerous options are available today that have potential for some measure of control, collectively speaking, they are limited considering the scope of the task of controlling so many genetically diverse insect forms on so many different crops and in so many different geographical areas and climates.

As is the case most often in agriculture, innovative practices, such as IPM, take time to gain acceptance and experience in the field. It is an evolutionary process and IPM is finding its way into conventional agricultural practices as modifications of the strategy are brought into accommodation for local conditions. Those who practice the plant sciences have been given a viable strategy for insect control but a little more information on the attending limitations is needed so it can be more effectively implemented in a wider array of practical situations.

As expected, since nature is not static but is evolutionary, no systems or means of insect control are without limitations. The extent to which limitations can be identified or predicted increases the value and likelihood of adjusting control measures to better assure success. Identifying limitations is where entomologists, plant, and food sciences become collaborators. Within the art and science of producing and processing food crops, there are both natural, physical, economic, and regulatory limitations caused by insects apart from those the entomologist faces in controlling the insects themselves. Unless these limitations are considered in the design and application of IPM, the level of control may be such that it can still leave the crop as unacceptable for harvesting and processing.

We must depend on entomologists to help design IPM for, say, spinach growers who have only about 40 days from planting to harvest, for corn growers who may have a 3 to 4 months interval, and for perennial crops that flower in the early spring and harvest in the late fall. I recall observing sweet corn processing operations reject corn infested with more than 10 to 12% corn ear worm. In some instances the corn wasn't even harvested because not enough help could be crowded on the processing lines (also occupationally dangerous) to remove the damaged ears and worms before entering the kernel cutters and containing the whole lot in excess of Food and Drugs standards for wholesome food. Broccoli excessively contaminated with aphids and worms shortly before harvest may have to be similarly rejected. Another example of inadequate control of insects creating a health hazard occurred in the southwest. Aflatoxin was found in milk in the stores. This was due to pink bollworm damage in cotton. Cotton seed is a high protein feed source and is fed to dairy cattle. The threshold level of insect control was set according to damage to bolls produced for the lint. A much lower food safety threshold level of infestation was needed to protect the cotton seed from *A. flavus* infestation.

We also need collaboration on the nature and level of insect control according to the obscure adverse effects insects have on the physiology, chemistry, and metabolism of food and feed plants in the field. It is known that some plants alter their metabolism to produce a chemical(s) not present in unstressed plants that may be toxicologically significant and capable of repelling feeding insects. In consideration of our legitimate concern for health and food safety, these metabolites may not be desirable chemicals to add to human and animal diets. A physiological dysfunction in soybeans caused by insect injury reaches beyond the obvious leaf damage to cause reduction in nitrogen fixation by the plant.

The complexities of agriculture and the interactions among the sciences and disciplines needed to address the unending problems naturally associated with food, feed, and fiber production have always been challenging by themselves and an interesting subject when observed in action. One such action in which entomologists are one of the principal players deserves to be mentioned in this essay as evidence that progress is constructive and underway. Entomologists are part of the planning group for the first of a series of International Congresses for the Implementation of Pest Resistance Management Practices to be held every three to four years until the objectives are met. The projects decided on will be practical and feasible. It is envisioned that the Congress will be the threshold for refining, integrating, and preserving all the existing elements of plant protection technology in the most agriculturally efficient and environmentally acceptable methods of application globally. It is long range; it is inevitable, and with patience and support it will succeed.

In looking in on entomology, and this will be part of the Congress, we also see the entomologists that have their niche in the discipline concerned with insects that debilitate animals and poultry, the veterinary and medical entomologists concerned with insect vectors of diseases. They have a difficult row to hoe considering that at least 80% of infectious diseases are insect borne and literally millions of people around the world die or are totally incapacitated by insect transmitted diseases. An almost forgotten entomologist, I see only fewer and fewer as I look in, is the insect taxonomist. Without this expert to correctly identify the insect that needs to be controlled, there may be misguided or canceled IPM programs. For those of us who have known the value of entomology and for nonentomologists on the outside looking in we would continue to hope for and welcome entomological collaboration, and willingness to recognize all the limitations to be met in plant protection, and we offer our help and support.

Policing Pesky Pests on Potatoes, Peas and Peppers

James J. Linduska and Joseph P. Linduska

Midway through 1988, *two* grapes of Chilean export were found to be contaminated by a highly toxic chemical. Immediately, all grapes on store shelves having that South American point of origin were rounded up and consigned to dumps. The action did little to quiet the public's hysteria and for some weeks thereafter grapes everywhere, irrespective of where grown, suffered a "bad press." Countless tons spoiled in produce markets.

But that furor was nothing compared to what occurred in February of the same year. Back then, a claim was made public that pesticide residues on produce would be responsible for the death of thousands of children, and one of the main sources mentioned was a compound widely used by apple growers. Responsible government agencies disputed the charge, yet it continued for weeks as a subject of mass coverage by most every newspaper, magazine, and TV network in the country. The incident caused a loss of $500 million to apple producers, in one state alone.

Happily for the Nation's food producers, over-blown affairs such as these are not of everyday occurrence; neither are they a source of genuine concern. But they illustrate well the preoccupation (if you want to call it that) of the American people towards matters of health and safety. And not human health alone, but that of most all living entities — fish, wildlife, forests — and the "health," as well, of the air around us, and water. It's all part of a growing environmental awareness. Some say it began with the Earth Day, 20 years past; some say in the last decade and others claim it's only now upon us.

For the entomologist, the precise date matters little. The fact is that people everywhere share a mounting concern about such things as tend to pollute the environment, whether it's smoke on airplanes or from unscrubbed stacks, the offal from our own living or the broad range of chemicals widely employed, and required, in modern-day living. And, one might add, this particularly concerns pesticides.

No one knows this better than the grower of vegetable crops, whether in the home garden or on commercial truck farms. The reason is obvious. The product resulting is expressly for human consumption and often without benefit of processing as might relieve contamination.

It's popular these days to speak of "challenges." Well, this one circumstance alone, pesticide toxicity, has been a full-blown challenge to most everyone involved in the production of vegetable crops. It's sent the chemists scurrying to concoct new compounds as would control given pests, yet meet the strict safety standards imposed by the Environmental Protection Agency (EPA) and the Food and Drug Administration (FDA). And it has placed great burdens on these regulatory groups to test a multitude of such products and find out just what those standards of safety should be.

The workload has filtered on down to the "end men," the field entomologists who test such new products under conditions of actual use to measure their efficacy on a variety of insects, on a variety of food plants and under a variety of working conditions. This continuing need for the development of new insecticides has been made necessary in a large measure by the remarkable ability of insects to develop "immunity" against toxicants.

In all living matter there is genetic variation among individuals which may lead to the survival (and change) of a species, even while most individuals of a generation or two may be killed off by some new factor in their environment.

For insects, a toxicant is such a "new factor." But while it may effectively destroy many individuals in one generation of a species, others are so constituted as to resist the chemical action and survive. Due largely to the enormous reproductive potential of insects, these few resistant ones beget many, and over the course of many (sometimes only a few) generations, a resistant strain results.

The Colorado potato beetle, a major pest of this plant, clearly shows what frustrations result for the control entomologist. As early as 1880, arsenicals were used to control this pest. In one form or another they continued effective up to 1940 when the first failure was noted. In 1945, the new wonder product, DDT, was employed and found to work well. But by 1952 signs of resistance were noted and by the following year there was a general failure of control effort. Over the past century, hundreds of chemicals have been tried in efforts to subdue this pest and dozens of widely different composition have been found effective. Yet, without exception, the beetle has developed resistance to one and all. With this insect, the evolutionary process leading to survival has outrun and outlasted the evolution of new chemicals to control it.

Development of new pesticides is a costly affair, and may entail a total outlay of $30 million before a new product results in sales. In addition to offering a useful level of toxicity to insects, they must likewise assure an acceptable level of safety to humans and other vertebrates. Then if their longterm use is foiled by uncooperative insects which, after a season or two, deny its toxic properties, sales of the product dry up. These two factors, cost of new product development (which in turn may be frustrated by insect resistance), and the growing public apprehension over broad spectrum toxicity, have led biologists to look for new alternatives. Out of this need was born Integrated Pest Management (IPM).

For entomologists the new procedure doesn't promise to eliminate completely the need for chemicals. However, it can be used profitably to *reduce* the need for toxicants (thereby extending their period of effectiveness), reduce the cost of insect control generally, and lessen the environmental shortcomings of widespread chemical use. IPM combines common sense with innovation and, in the days ahead, may be further strengthened by a host of all-new, and environmentally safe, control agents.

We can begin with the admission that in our complete dependance on chemicals we have over-used them, a shortcoming more appropriate to the home gardener, possibly, than to enlightened commercial growers. The new methodology (IPM) recognizes that not all infestations are of a degree to be of economic consequence. Accordingly, IPM begins there, by withholding the application of chemicals until surveys establish a true need. Fields are monitored and pest populations measured by a variety of methods. Only when infestations reach such predetermined levels as would lead to economic loss are toxicants applied.

This one procedure alone can produce enormous benefits, not only by greatly reducing hazards posed by toxicants, but in the dollar savings which result. In one California study, pest monitoring and deferred insecticide applications on fresh-market tomatoes, yielded savings of $84-$120 per acre without a decrease in yield or quality.

Studies in several other areas have likewise shown that insecticide applications may be reduced by as much as 50 percent without impairing quality or yield of some vegetable crops — and at substantial savings. So persuasive are the benefits, that in California most vegetable growers now follow such IPM practices, and it's rapidly catching on everywhere.

Beyond curtailment of pesticide use, the auxiliary IPM measures leading to control vary enormously, depending upon the crop involved, the specific pest and many other factors. Accordingly, we can only generalize here and mention *some* of the things effective for *some* species on *some* crops. For reasons which will be obvious, several are more appropriate to the home garden than they are to large commercial operations.

A day-dreaming entomologist once calculated that under optimum conditions and in a year's time, a pair of fleas could produce two *trillion* descendants, half the number of dollars in our national debt, and a single female of the common cockroach could give rise to two million young.

Such prolificacy is common in the insect world and suggests two things: one, "optimum" conditions for breeding are seldom met, and two, natural mortality is responsible for enormous losses during all stages of life; in total more than that occasioned by flea collars and aerosols. If that weren't so, insects would have long since inherited the earth.

Capitalizing on these two things: discouraging optimum breeding conditions and encouraging natural mortality, is basic to the IPM program. In some life stage, egg through adult, many crop pests overwinter in the soil and the litter of crops. Accordingly, sanitation as regards the disposal of crop residues is often important, and tillage practices may also be employed to allow weather and other natural forces to increase their toll. Crop rotation, in itself, has been a means to partially subdue some pests of farm and garden, and is a good procedure to follow in numerous situations. For example, successive plantings of squash on the same ground is a sure way to guarantee perpetuation of the troublesome squash vine borer. And the same may be said for other pests.

Plant breeders, alert to the problem of insects, have produced many varieties resistant to this major source of loss. Some new vegetable types repel insects; other are either unattractive or resist feeding. While not a total solution, they help and should be considered in planting programs. Neither will certified seed and pest-free transplants assure full relief, but they are more likely to lead to healthy plants which, in turn, are more resistant to insect damage.

Certain other types of cultural control, while not applicable to commercial growers, may be used to advantage in home gardens to further reduce the need for chemicals. Handpicking of adults and egg masses of larger, visible pests such as the Colorado potato beetle and tomato horn worm, can, in itself, control damage. The ravages of cutworms can be avoided by shielding young transplants with collars of cardboard or tin cans (with the ends cut out) and pressed into the soil. Shallow tins of beer set into the soil will attract and drown slugs. Direct spraying with a hose will dislodge aphids and mites, and the moist environment resulting will help avoid future mite build-up. Other insect pests of gardens (notably squash bugs) congregate in great numbers in hiding places. Boards placed on the ground will lure them in hordes where they can then be destroyed.

Aside from such physical factors as erode insect numbers, many biological agents give further check to their spread. A vast array of parasites, predators, and pathogens find in insects a source of food or a means for perpetuating their own kind. In common with other population restraints, they help to curb these pests but not many control them — at least to the degree required by commercial interests.

An organic farmer, or an informed and diligent home gardener, by utilizing all of Nature's decimating influences, could grow useful produce without benefit of chemical controls. Yet, even this much carries two assumptions: one, that his purist instincts would permit the occasional use of some botanicals and microbials, and two, that his pride in home-grown produce would cause him to overlook the puncture marks on a tomato put there by an errant stinkbug.

Commercial growers lack such latitude. The cosmetically sensitive American housewife wants produce that not only tastes good but looks good. Any sign of "nibbling by bugs" leaves vegetables spoiling on the store shelves. Worse yet, the Food and Drug Administration has what practically amounts to zero tolerance levels on contaminants such as insect parts in processed food — and it makes no difference if they're beneficial insects!

Therein lies a dilemma such as tends to shackle any all-out effort at IPM. Use sparing amounts of insecticides to satisfy environmental concerns and the result may be scarred vegetables, which the housewife won't buy and the FDA says you can't can. Yet, if you use too much you face difficulties stemming from Federal regulations limiting the permissible amounts of pesticide residues on items of food. Clearly, IPM needs additional allies in its war on insects.

Chemical pesticides for the most part are broad spectrum, that is, they are toxic not only to insects other than the target species but (in many cases) to humans and other vertebrates as well. That presents a problem, the obvious solution to which would be a product which could control the one pest you want controlled and just that one. An answer, at least a partial one, would appear to lie in microbials (biopesticides).

You recall, of course, the old jingle: "Big fleas have little fleas upon their backs to bite 'em, and little fleas have lesser fleas and so, ad infinitum." That truism applies to insects, no one of which is free of disease-killing pathogens which occur naturally and coexist with their insect hosts. Most often the relationship is fairly specific with only one kind of disease organism associating with one or a few kinds (species) of insects.

It's in this area that today's new hopes for safer insect control lies — with bacteria, fungi, viruses, nematodes, protozoa, and rickettsia, listed in descending order of probable utility. It's not so much a "new hope" as a resurrected one, given new life by ever-increasing environmental concerns and the commonplace result of insect resistance to chemicals.

While well over 1,000 microbials have been identified, the organism figuring most prominently in today's quest for microbial control is *Bacillus thuringiensis* (B.t.). As far back as the early 1900s it was found in Japanese silkworms and has been commercially available since 1939 when it first found a market in France.

More than 20 varieties of B.t. strains have been isolated from nature and used successfully against a variety of pests. In common with most insecticides of this type, they must be ingested to be effective, a drawback when compared with agrochemicals which are mainly nerve poisons and can be absorbed in a variety of ways. B.t. exercises its lethal effect through a spore-formed crystalline toxin which paralyzes the mouth parts and digestive tract and causes destruction of cells in the mid-gut. Starvation ensues.

The number of biopesticides which have reached commercial levels and are available for use on *all* insect pests is well over a dozen, but fewer than half of these are useful on vegetable crops.

Research on microbials has also teamed up with the fast-moving science of genetics with intriguing prospects resulting. Through the highly complex process of gene splicing some plants have been endowed with the capability of manufacturing their own insecticide and the effectiveness of one insecticidal microbe has been greatly enhanced through gene transfer to a different species of bacterium. As for other directions such research may be taking, it's hard to say. For competitive reasons, new product development is not a highly-aired subject.

What *is* known, is that for agrochemicals, research and development costs leading to a new discovery run about $20 million, plus another $5 million for toxicological tests. The annual sales required to recoup such investment costs are about $40 million. That calls for emphasis on pests of major crops.

Microbial control agents, however, are often easy to locate and the "discovery" outlay ranges well under 10 percent of the above. So too, the sales leading to profitability need be only 5 percent that of a new chemical insecticide. This much, in itself, would recommend vegetable-crop-insect-control as a profitable niche for research on microbials.

High cost aside, research chemists, likewise, are giving increased attention to development of new compounds as may overcome the problem of insect resistance and also meet the need for greater compatibility with environmental values. Two interesting developments concern insect growth regulators (IGRs) and synthetic pyrethroids. The former operates by interrupting the normal life cycle of the insect; the latter are synthesized chemicals, similar to, but not identical with, the natural pythrethrins. They promise many of the desirable qualities of the botanical product (fast knockdown powers and relative safety to vertebrates) and further add greater stability, toxicity and persistence.

The years ahead are likely to add useful weapons to the insect control arsenal of entomologists working with vegetable crops. It's doubtful that any will be of such a caliber as to allow total abandonment of chemicals or the dismantling of integrated pest management. Both of these will continue as necessary ingredients to the production of healthful foods in increasing quantity.

Two things appear certain: we can look for no surcease on the part of insects in their determination to survive and prosper; and the concern of humanity over environmental decay and degradation is likely to reach new highs, bringing greater demands for product safety. This much alone should be enough to insure vitality for the profession.

Bollworms

Dial F. Martin

The bollworm, *Heliothis zea* (Boddie), and the tobacco budworm, *H. virescens* (F.), often collectively referred to as the *Heliothis* spp. complex or "bollworms," are serious pests of cotton in the United States. Cotton may be attacked by either one or both species. The loss in crop damage and control in the United States from attack by the *Heliothis* spp. complex is estimated to be more than one billion dollars annually. These pests are members of the order Lepidoptera and family Noctuidae.

They are known to attack more than 70 kinds of plants including wild and cultivated groups. Some cultivated crops include cotton, corn, tomatoes, soybeans, and cowpeas. The bollworm is also known as the corn earworm, tomato fruit worm, vetch worm, and others, depending on the crop it attacks. It is a serious pest of sweet corn. The tobacco budworm does not normally attack corn, but is a long-standing pest of tobacco. It was not reported as a pest on cotton until about 1936. Both species are now major pests of cotton. The *Heliothis* spp. complex are general feeders attacking many kinds of plants. They feed in the fruits of plants such as corn ears, flower buds (squares) and bolls of cotton, fruits of tomatoes, and pods of beans and peas. They may feed on tender foliage and terminal buds of certain plants but prefer the seeds or fruit. When conditions are ideal, huge populations may develop that can destroy a crop.

Nature of Damage

The caterpillars or larvae of both species of *Heliothis* generally feed on the fruits, damaging or destroying the fruit of the plants on which they are feeding. However, during the early stages of plant growth before fruits appear, larvae feed on the tender terminal growth of the host plant. In the case of plants with a fruiting nature of the cotton plant, eggs are generally laid on the tender terminal growth and the newly hatched larvae start feeding on this tender foliage and may feed here for a day or two. Then they move to progressively larger fruiting forms causing the squares to turn brown and fall from the plant. In its journey down the plant, it will usually consume or damage several fruiting forms. Once a fruiting form is attacked by a larva (except for corn ears) it usually results in a total loss due to the squares "blasting" or entrance of microorganisms into bolls that cause it to rot. Both foliage and seed pods on alfalfa, vetch, and similar hosts may be attacked.

Description

The four stages in the life cycle are: egg, larva or caterpillar, pupa, and adult or moth. The moth's wing spread varies from 3 to 5 cm and varies in color from a light brown with a greenish cast in the males to a deep reddish brown in the female. The tobacco budworm adults are much like the bollworms except they have three oblique white lines across the front wings which distinguishes them from the bollworm. The two species are very similar in appearance. Moths feed on nectar of flowers and other sweet liquids. They are strong fliers and may migrate several miles in one night. Activity begins at dusk with egg-laying, feeding, and mating. Eggs are small waxy white objects, 0.5 mm in diameter with a flat base, dome-shaped, and ribbed. Mature larvae are about 3.8 cm in length, variable in color ranging from very dark to light green through rose and brown to almost black. Their bodies are usually marked by alternating longitudinal dark and light stripes but this is not always the case. The brown pupal stages (about 14-23 mm in length) are found in the soil 2.5 to 10 cm deep. The moths and immature stages of the *Heliothis* spp. complex are so similar that usually expert assistance is needed to identify the different species.

Life History

Moths prefer rapid growing succulent cotton or other crops for egg laying. Eggs are laid singly on host plants of the caterpillars. A moth may deposit as many as 3,000 eggs but the average is about 1000. During summer, eggs hatch in two to three days but may take longer in the cooler weather of spring and fall. The length of the larval stage varies from about two weeks in the warm temperature of summer to a month or more during spring and fall. The pupal stage varies in length from less than two weeks in summer to six months or more during diapause (period of hibernation or over-wintering). The bollworms over-winter as pupae as far south as southern Florida and the Lower Rio Grande Valley of Texas. Moth emergence in the spring may occur over a period of one month or more.

In most years there is a low level of development even during winter in the Lower Rio Grande Valley of Texas, southern Louisiana, and Florida. A few eggs may be found in January in southern Texas and Florida.

The first eggs are found in March in central Texas and April in central Mississippi and Arkansas. The first generation of bollworms develop on various weed and legume hosts such as Indian paint brush, bluebonnets, phlox, and many clovers. Cannibalism may occur on plants with over-crowded populations of the insects but the intensity of cannibalism is probably greater in the bollworm than the tobacco budworm. Because of slow growth during cool spring weather, cannibalism and the fact that larvae are less protected on these early hosts, parasites and predators take a heavy toll of the population. The bollworm moth usually emerges and becomes active in the spring before the tobacco budworm.

The preferred cultivated host of the bollworm is silking corn. When corn silks dry up and wild host mature, moths migrate to cotton and other cultivated crops such as grain sorghum, tomatoes, and soybeans. The tobacco budworm does not attack corn and grain sorghums but does feed on many of the same wild host plants as the bollworm. It is usually the third generation that migrates to cotton. In fall, some of the favorite crops for both species are regrowth cotton, soybeans, and alfalfa and fall wild hosts where over-wintering populations are produced. A generation may be completed in one month under warm conditions of the summer but may be extended over a period of two months in cold weather of spring and fall. There are four to seven generations yearly in the southern states but only one or two generations in the insects' northern range. Generations as a general rule are fairly well defined but toward the end of the season, there may be considerable overlapping and all stages may be found at one time.

Control of Bollworms in Cotton

For effective management of bollworms in cotton, a grower must follow closely the development of the *Heliothis* spp. complex population in early wild and cultivated host plants to know when migration to cotton occurs. Then a grower must be ready to apply effective chemical control measures if economic populations are indicated. Timing of control applications is of paramount importance. Control is much easier when eggs are hatching and worms are small. Large worms are difficult to kill.

Experienced growers cannot rely upon pesticides alone for control of bollworms because of development of resistance to pesticides. Pesticide use especially in early season for other pests tends to reduce parasite and predator populations that normally would reduce or hold bollworm populations in check. Other factors that favor bollworm population increases are large acreages of single crops such as cotton, corn, and other favorable host plants that make it easy for moths to find suitable egg-laying sites and larvae easy access to food. Improper use of irrigation and fertilizer add to the problem. However, research results show that with proper cultural practices coupled with proper use of irrigation and fertilizers, resistant varieties, short season cottons, management of parasites, predators, and diseases, and judicious use of pesticides, when required, aid in effectively reducing bollworm losses. Additional new techniques such as use of pheromones, sterile hybrids, and changes in cotton culture are being investigated and no doubt will contribute to better population management in the future.

Geographical areas of cotton production in the United States are many and varied because of many differences in climate, soil type, and agroecosystems. Because of these differences in areas of cotton production, great efforts have been made to develop effective integrated pest management programs for each area or even sub-areas. Improvements are continually being made in programs. Integrated pest management programs and guides for control of cotton insects are available from Agricultural Extension Services. These programs and guides for specific areas should be consulted for latest recommendations on integrated pest management practices and insecticides to be used for effective and economical control by the grower.

Suggested Further Reading

Brazzel, James,R., *et al*. 1953. Bollworm and tobacco budworm as cotton pest in Louisiana and Arkansas. Louisiana Technical Bulletin No. 482.

Ewing, K.P. 1952. The bollworm. *In* Insects, the yearbook of Agriculture 1952. pp. 511-514. Stefferud, (ed.). United States Printing Office, Washington, D.C.

Little, V.A. 1972. General and Applied Entomology. 3rd edition. pp. 289-291. Harper and Brothers, New York, N.Y.

Little, V.A. and D.F. Martin. 1942. The cotton bollworm. pp. 40-51. *In* Cotton Insects of the United States. Burgess Publishing Co., Minneapolis, Minn.

Southern Cooperative Series Bulletin 169. 1972. Distribution and abundance and control of *Heliothis* species in cotton and other host plants. Oklahoma Experiment Station, Okla. State Univ., Stillwater, Okla.

Southern Cooperative Series Bulletin 316. 1986. Theory and tactics of *Heliothis* populations management: I- cultural and biological control. Agriculture Experiment Station, Div. of Agric., Okla. State Univ., Stillwater, Okla.

Southern Cooperative Series Bulletin 337. 1988. Theory and tactics of *Heliothis* population Management: III- emerging control tactics and techniques. Agricultural Experiment Station, Div. of Agric., Okla. State Univ., Stillwater, Okla.

The Boll Weevil — Lookin' for a Home

Theodore B. Davich

Boll weevils were first found in Veracruz, Mexico, in 1830 by L. A. A. Chevrolat. He sent the specimen(s) to C. H. Boheman, a systematic zoologist in Sweden, who described the species and named it *Anthonomus grandis*. It is reported to have crossed the Rio Grande in 1892 near Brownsville, Texas. By 1922 it blanketed the entire cotton growing area in the Southeastern U.S.A. and much of Texas. It was first reported from Arizona in 1920, from Venezuela in 1949, from Columbia in 1951, California in 1982, and Brazil in 1983.

The boll weevil has 4 life stages: the egg, larva, pupa, and adult. The adult lays its eggs in the floral bud, commonly called a "square," and in small bolls. After the eggs hatch the larvae feed on the surrounding tissue causing virtually all of the squares and occasionally some of the small bolls to drop to the ground. The larva then completes its development to the pupal stage followed by transformation to the adult stage. The adult then breaks its way out of the dropped squares and bolls and climbs up a nearby cotton stalk and within 3 to 5 days mates to begin the cycle again. The length of the cycle varies with temperature. Under favorable temperature conditions the cycle could be about 16 days to as long as about 50 days during the cooler days of the fall.

The boll weevil created disaster almost everywhere it was found. During its spread from Texas to North Carolina, it caused tremendous losses to cotton. At the time of its spread in the Southeastern U.S.A. the region was a one crop economy. Financial losses were tremendous. It was not uncommon for banks, businesses, farmers, doctors, and others to go into bankruptcy.

Shortly after the boll weevil was first found near Brownsville, Texas, consideration was given to the idea of preventing its further spread, or even eradicating it, by establishing non-cotton zones. In 1895, Dr. C. W. Dabney, Assistant Secretary of the USDA appeared before the Texas Legislature urging enactment, unsuccessfully, of a law to create non-cotton zones to prevent further spread of the boll weevil into the Cotton Belt. In 1903 the Texas Legislature offered a reward of $50,000 to anyone who could find a way to eradicate the boll weevil. In 1903, a plan was promoted, unsuccessfully, to establish a non-cotton belt along Louisiana's western boundary to prevent further spread of the boll weevil. Governor Huey Long in late 1931 had the Louisiana legislature pass into law a bill to prohibit the growing of cotton in the entire state during 1932. Because other southern states did not follow suit the law was rescinded in 1932.

Insecticides were advocated for control shortly after the weevil appeared in the U.S.A. Paris green, an arsenical compound, probably was the first one tried. It turned out to be ineffective and phytotoxic as well. Lead arsenate was tested and found to give inconsistent results. Calcium arsenate proved to be effective and relatively inexpensive, but had major side effects. It resulted in the development of high and frequently damaging infestations of the cotton aphid. Also, it resulted in phytotoxicity to legumes planted the year following treated cotton.

Extensive use of organic insecticides began at the end of WW II with the introduction of synthetic organochlorine compounds such as benzene hexachloride, aldrin, toxaphene, dieldrin, endrin, DDT, and others. However, by the mid-fifties the boll weevil had developed resistance to these insecticides. This was first reported from Louisiana in 1955 and soon became widespread throughout the Cotton Belt. The organophosphorous compounds came into general use following the development of resistance to the organochlorine insecticides. Among these were methyl parathion, azinphosmethyl, malathion, and others. Systemic insecticides, those that are translocated by the cotton plant to the site of action, were extensively tested during the late-fifties to the mid-sixties. These included aldicarb, thimet, dimethoate, and a host of experimental compounds known only by company number. Research was directed towards finding a systemic insecticide that could be applied to the seed before planting, as well as in-furrow or sidedress applications, or to the plant. Boll weevil control was generally poor to fair under field conditions and some compounds resulted in stand reduction or favored the buildup of bollworms.

Some carbamate insecticides were found to be effective, but did not become widely used because of high mammalian toxicity or high cost.

Crops planted prior to the normal planting date, generally along field edges, were used to lure overwintered weevils to them for treatments with insecticides. In these so called "trap crops" it is important to achieve maturity differential between the trap crop and the regular crop for greatest effectiveness. In some areas this has been done easily and with some degree of success. In many others it was not adopted. The necessity for earlier planting of the trap crop, or planting seed varieties, with two different early fruiting characteristics, has tended to inhibit wide acceptance by individual growers. Some areas delay planting in order to deny the weevil food and oviposition sites, resulting in suicidal emergence. This strategy might be compromised by the trap-crop system.

A nematode parasite of the boll weevil adult was discovered in 1961, but was ineffective in causing widespread reductions in weevil populations. The kelep ant, predacious on boll weevil larvae in squares, was imported from Guatemala and found to be ineffective and failed to survive U.S.A. winters. A parasitic wasp, imported from Africa, was studied in the mid-thirties and research on it was repeated and expanded in the mid-sixties. Both attempts failed to provide adequate control and the wasp did not become established in the U.S.A. Various diseases of the boll weevil have been found, but none proved to have any control value or were insufficiently researched to establish their value in control efforts.

Attempts to develop physical methods for control were tried shortly before the turn of the century. These included hand picking of adult weevils and infested shed squares on the ground, shaking plants by hand or by a machine and collecting the dislodged weevils and infested squares in pans, burning trash at gins or cotton plants after harvest, and the use of light traps. These measures proved to be marginally effective or totally ineffective. In the mid-sixties a number of mechanical devices were built and tested for destruction or collection of weevils and infested squares. A flail machine was designed and tested to destroy infested shed squares. It competed successfully with insecticides until migration into the test fields began. However, it posed a potential hazard to property, humans, and animals. A vacuum device for removal of infested shed squares was tested in field cages. A flame cultivator, originally designed for weed control, was modified by extending the hooded burners to 34 inches and tested against shed infested squares. It performed fairly well, but the necessary low speed of one-half mile per hour precluded its adoption as a control method. A device utilizing hoops of chains was used to drag infested shed squares into the row middles where exposure to the sun would kill immature boll weevils. Economics and hazards to farm workers and property prevented the adoption for general use of the only promising devise, the flail machine.

The chemosterilants, hempa and apholate, applied as sprays directly to cotton, were researched in the mid-sixties. Although the results of such tests, conducted in large cages, showed promise, further research was abandoned because of the mutagenic properties of the known chemosterilants. No further research on this control method is known to have been conducted.

Baits, containing insecticides or other lethal agents, were extensively tested for boll weevil control. At the turn of the century a bait containing water, molasses, and an arsenical was tried and found to be ineffective. Two protozoan pathogens also were tried in baits containing boll weevil feeding stimulants found in cottonseed oil. Although effective control was achieved, problems with formulating the bait and its short shelf life were major obstacles to its further development.

Crosses of the boll weevil with various "sister" species were attempted in order to produce a sexually active, but sterile hybrid. The level of sterility obtained was either too low to be of value in a release program, or physiological problems developed that precluded further research.

Attractants, repellents, and antifeeding compounds were found to be present in cotton or in related malvaceous plants. Research on these compounds failed to provide new approaches to control.

The cotton industry, in the 1950s requested from Congress a large increase in funding for research and facility needs to meet the problem. Congress appropriated, in 1959, over 1 million dollars for the construction of a Boll Weevil Research Laboratory which was located adjacent to the campus of Mississippi State University. In addition, substantial increases in funds were allocated to several other existing federal laboratories and state agricultural experiment stations.

The expanded research findings that resulted from these expenditures led the cotton industry and federal and state agencies to devise a plan to determine whether or not the weevil could be eradicated from the U.S.A.

A large scale experiment, entitled the Pilot Boll Weevil Eradication Experiment, was conducted in South Mississippi and parts of adjoining Louisiana and Alabama. In the early 1970s a Technical Committee and a National Science Foundation Committee both concluded that it was technically and operationally unfeasible to eliminate the boll weevil as an economic pest from the U.S.A. However a significant number of entomologists thought eradication could be achieved.

Diapause in the boll weevil is characterized by a very high increase in body fat, atrophy of the reproductive systems and elimination of free water. This physiological state permits the weevil to survive U.S.A. winters and begins to manifest itself in late summer and fall. Insecticide treatments aimed at the onset of diapause will greatly reduce populations that would survive to infest the next seasons crop. These treatments have been termed diapause control.

The components used in the Pilot Boll Weevil Eradication Experiment were diapause control, insecticides, Grandlure in traps, release of sexually sterilized males, and quarantine measures.

Recommended boll weevil insecticides for the Pilot Experiment were used in diapause control and early in the next season. Then

Grandlure, emitted by males to attract both sexes was used in traps. Grandlure was isolated, identified, and synthesized by chemists and entomologists in the mid sixties and was placed in traps especially designed to trap and retain mainly females.

Research on rearing, sexually sterilizing, and releasing millions of boll weevils was accomplished in the late 1960s. Quarantine lines were located to prevent weevils in infested cotton from moving, by farm equipment and other vehicles, into the Pilot Experiment. These measures were used for the second and third years and was followed by an extensive trapping program to detect any small loci of weevils that would then be eliminated with the extensive use of traps and the judicious use of insecticides.

The apparent, though debated, success of eradication in the Pilot Experiment, prompted the cotton industry, federal and state agencies to attempt a trial eradication from the entire cotton belt. The selected starting area were cotton fields in Virginia and part of North Carolina. The program started on about 64,000 acres in 1978.

With eradication achieved in this Boll Weevil Eradication Trial area, additional segments were added in a southern direction until all of Virginia, North Carolina, South Carolina, Georgia, Florida, and the Southern counties of Alabama were included in the program by 1987. The present Boll Weevil Eradication program was modified, because of opposition by environmental groups and a number of growers. They insisted that the U.S. Department of Agriculture develop an Environmental Impact Statement. Once the statement is issued and accepted, further eradication increments can be included in the program. The measures used in the program are malathion treatments for diapause control, malathion, Grandlure baited traps, at the rate of one per acre and quarantine, or buffer lines.

Northwestern Mexico, Southeastern California and Western Arizona eradication is being achieved, also with malathion, traps, and quarantine lines or buffer zones. This effort started in 1985 and does not utilize diapause control at the beginning of the program.

Research is continuing on control measures that can be adopted to fit into an eradication program. Some of these include an intensified research program to develop cotton varieties resistant to the boll weevil. In addition to the classical methods to develop resistant varieties genetic engineering is being researched.

A renewed interest in boll weevil diseases, used in conjunction with feeding stimulants and attractants, is being studied. Recent breakthroughs in formulating Grandlure with insecticides and feeding stimulants in a plastic bait have provided a promising control method. The bait device, in the form of a coated stake use only a fraction of the normal insecticide per unit area as compared with conventional applications and spares beneficial insects to control secondary pests. The measures used in the areas were insecticide treatments with malathion, Grandlure baited traps, and quarantine lines. In the western states, part of California, Arizona to New Mexico eradication was achieved also with malathion, traps, and quarantine lines.

Insect Pests
Of Deciduous Fruits
In the Mid-Atlantic States

Arthur M. Agnello

It is perhaps difficult for many of us to appreciate the complicated process undertaken to assure a continual supply of high quality fresh fruits and processed fruit products to our supermarkets. Not so long ago, most fruits of any kind were considered luxury items by the average consumer, but the fruit industry has taken great strides in the last few generations, and the insect pests of fruit have grown with it. Nothing illustrates this process better than the history of apple production in New York State, which was well under way by the time of this nation's formal beginnings. Missionaries and early settlers in the mid-1700s customarily planted apple trees from seeds or other propagation stock as a natural component of settling the land, and by the middle of the nineteenth century, a small apple orchard was an integral part of every landowner's property. It provided to the farm not only fresh fruit, but also a large array of apple products that were dried, preserved, pressed, fermented, and distilled. A census in 1875 estimated well over 18 million apple trees in the state, and although today's count of cultivated trees is only one-fourth that number, wild or naturalized trees certainly far outnumber the domesticated ones, most having been planted by the inadvertent feeding and dispersal habits of farm animals such as the dairy cow. As a consequence, there has been ample opportunity for the evolution of entire food systems based on the biological relationships between apple and the organisms associated with it, such as insects and mites.

Early researchers in New York estimated nearly 500 species of insects known to feed on apple; fortunately, only a relatively small number of these ever reach economic pest status. A cooperative survey conducted in 1983-84 identified a total of 191 phytophagous (plant-feeding) insect species in Virginia, West Virginia, Pennsylvania, and New York in managed and abandoned apple orchards. Most numerous were species of Lepidoptera (mostly moths and their caterpillars, 43%) and Homoptera (leafhoppers, plant bugs, aphids, and scale insects, 32%). Current tabulations of actual economic pests list just over 60 species in New York, approximately half of which are considered important enough to warrant specific control recommendations in the most widely used university and industry production guides. This compares with 17 pest species for pear, 12 for peach, and seven each for tart cherry and plum. A brief examination of any crop's history will reveal the close relationship between its key pest problems at a given time and the pest control efforts that have been used in the past. In the 1983-84 survey cited previously, managed orchards were found to contain fewer phytophagous species than did abandoned orchards, although a greater proportion of those present in the managed sites were pests. Accordingly, the species currently responsible for the majority of pest management decisions on modern apple farms vary to some degree from the worst offenders of 80, 50, or even 20 years ago. Although some of this change has to do with extraneous factors — developmental changes in the rural areas, modifications of planting techniques, and subtle ecosystem shifts — most of it is a direct result of chemical pesticide use, and represents responses of insect and mite populations to long-term exposure to various toxicants. In too many cases, this has taken the form of an acquired tolerance or resistance to certain pesticides by the target species, coupled with deleterious side effects on non-target species, such as destruction of natural enemies and induction of secondary pest infestations. This pattern has been repeated in most crop systems.

Although New York's apple tree population reached a record high around 1875, the planting of commercial orchards was generally in decline within the next 10-15 years, simply because of grower discouragement at the increasingly heavy crop losses from insect pests and diseases. This caused enough concern among agriculturists to stimulate the formulation and implementation of some of the first effective pest control measures, which were both cultural and chemical in nature. By the turn of the century, tree fruit production had stabilized into a specialized branch of the state's agriculture. In 1908, however, the entire apple industry was threatened by the first documented case of insect resistance to an insecticide, which was that of San Jose scale to lime sulfur. Even so, the incidence of resistance among insects and mites of apple was of minor importance until lead arsenate began to be used extensively (1920-1940), and was restricted to the relatively few pests constantly being exposed (such as codling moth). Economic factors, aided by the extensive chemical developments generated during World War II, then initiated what was to become a routine search for effective new classes of compounds to use against apple pests that had become resistant to the last ones. The new products were always relatively unselective, and each initially successful introduction was eventually followed by a round of control failure or decline, or problem infestation by some previously innocuous species. Thus, resistance was recorded by codling moth to lead arsenate in the 1930s and to DDT in 1951; by redbanded leafroller to TDE in 1954; by spider mites to parathion and malathion in 1955; by European red mites to many compounds in the 1950s; by aphids to organophosphates in the late 1950s; by leafminers to organophosphates in 1976; and by white apple leafhopper to DDT in 1959 and to organophosphates in 1970. It is notable that the most problematic pests in this region's apple production systems today are generally not the fruit feeders that were once such serious threats — codling moth, apple maggot and plum curculio — but rather the indirect, foliage feeders that were formerly regarded as secondary pests, if they occurred at all. These include leafminers, leafrollers, leafhoppers, and plant-feeding mites.

Some may be inclined to view this situation and inquire whether we weren't better off 75 years ago, when there were only one or two serious pests troubling the crop, and few pesticide sprays applied in return for a "tolerable" amount of damage. Leaving aside for a moment the feasibility of reestablishing such conditions, it is instructive to first consider the inherent ability of an apple (or peach, pear, plum, or cherry) to sustain itself in a typical temperate growing region. It is true that wild apple fruits often display an unexpected lack of insect and disease damage; for instance, the codling moth seldom infests 20% of a tree's crop, and may not even be present at all in some remote stands. However, the pests to which the fruit do succumb are frequently devastating, and often as a direct result of the varietal qualities that make the fruit marketable. Beginning in 1960, a 1-acre planting of McIntosh, Macoun, and Cortland trees in western New York, once part of a commercial orchard, was maintained normally for a 10-yr period except for the omission of all insect and mite control spray treatments. European red mite, an otherwise serious problem on most apple farms, did not reach pest levels, but codling moth, plum curculio, redbanded leafroller, and apple maggot rendered the crop commercially worthless after the first year. The apple maggot was consistently the most damaging pest attacking more than 75% of the fruits each year after the second non-insecticidal season. In addition, a little-known pest in commercial sites, the lesser appleworm, caused severe fruit injury during the final two seasons. A similar outcome ensued in a 1984 New York study comparing insect and mite damage in standard and disease-resistant apple plantings that were left unsprayed by conventional pesticides for four years. In general, the disease-resistant cultivars were as susceptible to the various arthropod pests as were the standard cultivars. Canadian researchers assessing codling moth management techniques in 1975 found that, after only one season in which orchards were left unsprayed, a resultant 5% infestation level at harvest time translated into a 50% infestation level the following season.

The recurring presence in all these studies of pests from three major (and therefore distinct) insect orders — Lepidoptera (caterpillars), Coleoptera (beetles), and Diptera (flies) — is indicative of the formidable complex of pest species attacking tree fruits in the northeastern part of the country, and the difficulty of reducing them to acceptable levels. Furthermore, the majority of these species are not restricted to a single host fruit. The codling moth, originally a native of Eurasia, is one of the most cosmopolitan tree fruit pests. Its larva, the traditional "worm" of the apple, attacks the fruit of a wide range of plants, including apple, pear, quince, crab apple, and walnut. It may have as many as 3 1/2 generations per year, depending on the locality and the length of the growing season. A closely related species, the oriental fruit moth, is most important as a pest of peach, but is commonly found on nearly all deciduous fruit trees. The plum curculio, a native of North America, is a snout weevil that is a common pest in virtually all commercial apple orchards, although it also attacks plums, nectarine, apricot, cherry, peach, pear, and quince. The adults feed on and oviposit in the fruit as soon as it starts to form, continuing until the apples are about 1 1/2 inches in diameter. The larvae cannot complete development in hard fruits such as apples or pears that stay on the tree until harvest maturity, but unsightly scars remain on the fruit surface as evidence of their activity. The San Jose scale, originally from China and introduced into California during the last century, attacks not only apple, but also pear, peach, plum, and sweet cherry. Adult scale infestations on the bark contribute to an overall decline in tree vigor, growth, and productivity. Fruit feeding causes distinct red-purple spots that decrease the cosmetic appeal of the crop. Leafrolling caterpillars such as the obliquebanded leafroller feed on both the fruit and foliage of apple, pear and peach. Even the apple maggot, known primarily as the most important pest of apple in New York, can be problematic in plums, as is part of a species complex that infests cherries and blueberries.

Despite their relative abundance, tree fruits are still regarded as something of a luxury commodity by the average consumer, and great commercial value is placed on their cosmetic appeal, which means that essentially nothing less than a perfect looking fruit is acceptable to most buyers. With so many insect and mite pests capable of despoiling the crop (not to mention diseases, birds, and other animals), it is not surprising that apples are among the most highly sprayed crop commodities on a per-acre basis. Including fungicide, miticide, and insecticide sprays, a New York apple orchard could conceivably need as many as 10-12 separate pesticide applications in a growing season, most of which would be combinations of two or more different materials. Notwithstanding the less positive aspects of over-reliance on chemical control measures, the proper use of synthetic pesticides has come to play an essential role in the development of improved pest control tactics, specifically within the context of integrated pest management (IPM) approaches. To be certain, pesticides are extremely important components for IPM systems for apple, by virtue of the development of more selective compounds, more effective formulations, and better application techniques. Pesticides are intended for use in contemporary IPM programs primarily as specific tools for specific tasks, and as a last recourse for control rather than as an initial response.

The earliest true implementation of this philosophy in apples resulted from efforts in Nova Scotia during the 1950s, which took advantage of the developed resistance in predatory mites to a commonly used class of apple insecticides, the organophosphates. By limiting insect control sprays to these materials as much as possible, growers were able to effect substantial natural control of pest mites through predation by the beneficial mites. This work stimulated further investigation into additional natural enemies that could be encouraged to survive in apple orchards by the use of selective chemicals. Consequently, today the beneficial insect and mite species in each regional production system are taken into account as much as possible when pesticide recommendations are made. In Pennsylvania, consideration for the mite-feeding ladybird beetle, *Stethorus punctum*, is central to the schedule and field configuration of spray applications made against most of that state's insect and disease pests.

Determining insect pest development and abundance for management purposes has become a standard element in most fruit IPM systems since the 1970s, when entomologists began identifying and synthesizing the pests' sex pheromones, which are messenger chemicals insects release into the air to facilitate mating. By incorporating dispensers of these chemical mixtures to simple traps, a grower can monitor to establish when a particular moth is first present in the orchard, when it reaches peak numbers, and use the catches to help predict the most appropriate period in the insect's life cycle to apply control measures. Work is still in progress on the technique of releasing sufficient pheromone quantities to disrupt mating communications of a number of moth pests and therefore prevent the production of their troublesome offspring. The feasibility of this technique has already been demonstrated for codling moth, redbanded leafroller, obliquebanded leafroller, and oriental fruit moth, but further refinements are needed to make it economically advantageous in more production systems.

Renewed investigative efforts into fruit pest biology have prompted advances in a range of strategies that are often combined in current pest management activities. The most fundamental, and therefore the most challenging of these involves prediction of pest development and population dynamics, once a nearly impossible task, but now more approachable with the help of computer simulation models. One underlying aspect of this endeavor has already exhibited its usefulness, because it is easily understood and implemented; this is the practice of tallying accumulated heat units by charting daily temperatures, to gauge the developmental progress of local populations. Closely related to this area because it is in fact a sub-category of pest modeling, is the elucidation of "action thresholds" — that level of pest occurrence high enough that the resultant crop damage would warrant some control measure. Such a pronounced departure from the previous zero-tolerance philosophy of pest control obviously relies on a detailed understanding of each insect's ecological importance in the crop system, which requires much research and field testing. The concept of routine crop "scouting" to assess pest levels is promoted as widely as is permitted by the often incomplete biological knowledge at hand, and certainly much improvement is needed. For some pests, such as apple maggot, behavioral responses to crop appearance and smell can be exploited in monitoring devices that reveal the insect's presence before it can damage the fruit. The efficiency of these traps is high enough that chemical sprays can be withheld until and unless the adult (a fly) is actuary caught. The integration of this large number of diverse elements into a pest management program for individual fruit growers could not occur without the availability of complex information delivery systems, including computer networks, recorded phone message services, newsletters, radio broadcasts, and other more traditional avenues of information extension. Large scale computer-assisted decision support systems are not far off.

Other future developments in the insects vs. fruit confrontation will sound exotic and impractical upon first consideration, but will follow naturally as we come to understand more of insects' fundamental nature. There is currently a federal program being undertaken to defeat the notorious Medfly in western states through mass releases of male flies made sterile by irradiation. New classes of non-toxic chemical compounds under development cripple insect biochemical processes by mimicking the hormones they themselves produce. Genetic manipulation of insect pests or their food plants seeks to short-circuit the conditions that make the host/pest associations possible. Such is the commitment to this contest over a portion of our food supply, that one might easily regard the fruits purchased as a matter of course with the rest of the groceries, as the spoils of a rather sophisticated battle of wits between two extremely clever groups of this world's inhabitants.

Suggested Further Reading

Agnello, A., J. Kovach, J. Nyrop, & H. Reissig. 1990. Simplified insect management program: a guide for apple sampling procedures in New York. Cornell Coop. Ext. IPM Bull. 201A.

Brown, M. W., C. R. L. Adler, & R. W. Weires. 1988. Insects associated with apple in the mid Atlantic states. New York Food Life Sci. Bull. 124.

Chapman, P. J., & S. E. Lienk. 1971. Tortricid fauna of apple in New York: including an account of apple's occurrence in the state, especially as a naturalized plant. Spec. Pub. New York State Agric. Expt. Sta., Cornell Univ., Geneva.

Croft, B. A., & S. C. Hoyt [eds.]. 1983. Integrated management of insect pests of pome and stone fruits. Wiley & Sons, New York. 454 pp.

Glass, E. H., & S. E. Leink. 1971. Apple insect and mite populations developing after discontinuance of insecticides: 10-year record. J. Econ. Entomol. 64: 23-26.

Reissig, W. H., R. W. Weires, G. C. Forshey, W. L. Roelofs, R. C. Lamb, & H. S. Aldwinckle. 1984. Insect management in disease-resistant dwarf and semi-dwarf apple trees. Environ. Entomol. 13: 1201-1207.

Slingerland, M. V., & C. R. Crosby. 1914. Manual of fruit insects. MacMillan, New York. 503 pp.

Tette, J. P., E. H. Glass, D. Bruno, & D. Way. 1979. New York tree fruit pest management project 1973-1978. New York Food Life Sci. Bull. 81.

Tette, J. P., J. Kovach, M. Schwarz, & D. Bruno. 1987. IPM in New York apple orchards development, demonstration, and adoption. New York Food Life Sci. Bull. 119.

Whalon, M. E., & B. A. Croft. 1984. Apple IPM implementation in North America. Ann. Rev. Entomol. 29: 435-470.

Interregional Research in Resolving Some Insect Pest Problems of the Corn Belt

William B. Showers

Sustainable agriculture (low inputs) should continue to gain momentum with corn producers. But an impediment to this concept in the Corn Belt is prophylactic planting-time treatment for a number of insect pests, including the cutworm complex (black cutworm, *Agrotis ipsilon*, claybacked cutworm, *A. gladiaria*, variegated cutworm, *Peridroma saucia*, and dingy cutworm, *Feltia ducens*, and the armyworm, *Pseudaletia unipuncta*. Reduced prophylactic pesticide applications for cutworm and armyworm will come only after influencing producers' perceptions of risk through precise notification of timing and location of potential outbreaks.

As recently as 1977, some scientists, and agriculturalists in general believed that black cutworm, variegated cutworm, and armyworm overwintered as various developmental forms within the Corn Belt. At that time a Regional Research Project funded by the Environmental Protection Agency (EPA), brought together a team of Experiment Station and USDA, Agricultural Research Service (ARS), entomologists, modelers, and economists under the leadership of Drs. M. L. Fairchild, University of Missouri and E. E. Ortman, Purdue University to research the "Development of Pest Management Strategies For Soil Insects on Corn". Drs. S. L. Clement, Ohio Agricultural Research and Development Center; F. T. Turpin, Purdue University; A. J. Keaster, University of Missouri; Z. B. Mayo, University of Nebraska and W. B. Showers, Corn Insects Research Laboratory, USDA/ARS-IOWA State University investigated the cutworm complex attacking corn in the Corn Belt.

D. W. Sherrod, a graduate student of Dr. W. H. Luckmann, at the University of Illinois, presented convincing evidence that the most destructive cutworm, the black cutworm, seemed to be in fields before corn was planted. Unplanted fields with previous year's debris (especially soybean debris) and tender spring weeds seemingly were more likely to develop a cutworm problem after corn planting than fields without these attributes. Oviposition preferences (debris and young weeds) of the female moths, determined by M. K. Busching and F. T. Turpin of Purdue, and small size of black cutworm and variegated cutworm larvae during April and May suggested to R. N. Story and A. J. Keaster, of the University of Missouri, that these damaging larvae had not necessarily overwintered in these problem fields. During this time, L. V. Kaster and W. B. Showers (USDA/ARS) of Iowa State University presented a substantial premise that numbers of moths captured in light traps or traps baited with synthetic sex pheromone of the black cutworm female increased during nights with strong winds from the south. Climatologists were then brought into the investigations.

During the mid-1980s a pilot test funded by USDA/ARS brought together a team of entomologists from Iowa, Louisiana, Missouri, and Texas with climatologists and an economist at Iowa State University. Thousands of black cutworm moths were internally marked with a lipid soluble dye (Calco Red, Drs. W. B. Showers and F. Whitford) and released from Louisiana and Texas (Drs. J. F. Robinson and J. D. Lopez). Releases were made only after climatologists (Drs. R. E. Smelser and S.E. Taylor) had forecasted the presence of weather systems that we hypothesized to be conducive for northward movement. Meanwhile, traps baited with sex pheromone of the black cutworm female were stationed 16 km apart across an expanse of 620 km east to west and at 108 km interval south to north from central Louisiana to the Iowa-Minnesota border. These traps were observed on 7-day intervals by Drs. A. J. Keaster (University of Missouri), J. F. Robinson (USDA/ARS and LSU Rice Station), W. B. Showers (USDA/ARS and Iowa State University), and their colleagues.

Surface synoptic weather maps, atmospheric trajectory forecasts, and observational upper-air reports (radiosound data) were used by Drs. R. B. Smelser and S. E. Taylor (Iowa State University) to estimate the relative displacement of moths. Table 1 presents the average numbers of marked black cutworm moths recaptured after long-range transport. These studies, and others by Dr. J. N. McNeil, University of Laval, Quebec, and Drs. G. D. Butin, L. P. Pedigo, and W.B. Showers, (USDA/ARS) Iowa State University, presented powerful confirmation that each spring black cutworm, armyworm, and variegated cutworm moths are capable of arriving in the corn belt in two to four nights. The atmospheric low-level jet (300-900-m elevation) and a surface synoptic system of low pressure over the Plains States that induces air flow northward seemingly are necessary for moth transport.

Information gathered from these studies are now being used to develop and refine dispersion models. These models will be integrated with information gathered on geographic boundaries of winter populations and biological and environmental cues necessary for moths to initiate and sustain long-range flight. Other data being incorporated by these interdisciplinary, interregional research teams are: (1) moth influx nights; (2) larval damage period (time after a specific influx of moths that offspring are old enough to damage corn plants); (3) crop susceptible period (emergence of corn plants until plant is large enough to survive damage); (4) cohort damage period (larval damage period superimposed on crop susceptible period); and (5) seasonal damage potential (total cohort damage periods during a specific spring). Hopefully, when completed, this research package will allow precise notification of timing and location of potential outbreaks of these immigrant pests. Inputs, therefore, for suppression of these insects could then be reduced and implemented as rescue treatments only when necessary.

Table 1. Summary of *Agrotis ipsilon* males available for release and the number recaptured after long-range transport, 1984-1985 (Showers *et al.* 1989. **Ecology** 70:987-992).

Sites	Dates	Number A. *ipsilon* males			Sites	Distance from release (km)	Observation dates
		Released	Flyable	Recaptured			
Crowley, Louisiana 1984	4-11 June	6938	*	1	11.2 km south east of Elk-hart, IA	1142	13 June
				1	12.8 km east of Centerville, IA	1043	15 June
	19-26 Aug.	17,000	11,881	0**			
College Station, Texas 1985	16-28 May	11,950	9,799	1	1.0 km north of Bates City, MO	960	23 May
				1	4.8 km west of Sac City, IA	1175	29 May
				1	32.0 km west of Kansas City KS	921	6 June
	8-15 June	15,000	12,150	0**			
	17-21 July	16,000	3,520	1	11.2 km west of Rock Rapids, IA	1266	24 July
Total Capture				6			

*   Data on flyable males were not collected.

** Winds from south collapsed during release period.

An Abbreviated History of Insecticide Toxicology

Robert L. Metcalf

The history of the massive use of synthetic insecticides dates from the discovery of the stomach poison insecticide paris green by an anonymous innovator in 1865. Lead arsenate was subsequently devised as a safer plant protectant, and with the cheaper calcium arsenate the use of these arsenical insecticides in the United States rapidly increased to a maximum of about 156 million pounds by 1943. Insecticide toxicology emerged as an important specialty of applied entomology because of concern about the human health effects and environmental fate of this large burden of highly toxic xenobiotics and because of basic scientific curiosity about how these arsenicals kill insect pests. Concern about toxic residues on apples was responsible for the development of the analytical chemistry of pesticide residues about 1916 and in 1927 the U.S. Food & Drug Administration set a tolerance level of approximately 3.6 ppm arsenic trioxide on fruits and vegetables. This was lowered to 1.4 ppm in 1932 and returned to the original level as various perceptions developed about the health hazards of arsenic and lead.

Sodium fluoride was patented as a stomach poison in England in 1896 and cryolite or sodium fluoaluminate was introduced in 1929 as a safer substitute for arsenicals for the control of the codling moth and other chewing insect pests. Its use reached a maximum of about 16 million pounds annually just prior to World War II.

Other insecticides used in applied entomology during the first half of the last 100 years included the plant products nicotine, pyrethrum, and rotenone. As knowledge of basic insect physiology and biochemistry developed, pioneering insecticide toxicologists sought to account for their toxic action on insect organ systems, *e.g.*, the heart, the gut, and the nervous system. During this period quantitative toxicology appeared with emphasis on the dosage-mortality curve and the determination of the LD_{50}. The introduction of probit analysis and of innovative evaluation technics such as the leaf-sandwich method, the spray tower, and later of precision topical application; defined insecticide toxicology as a quantitative science and led to structure/activity studies of various insecticide types and to more rigorous studies of mode of action.

The dinitrophenol insecticides originated in Germany in 1892 when potassium 3,5-dinitro-o-cresylate was patented as a dormant spray for fruit trees. This development presaged the era of synthetic organic insecticides and extensive studies of structure/activity were made during the period of 1920s and 1930s. These investigations together with comprehensive studies of the cyclohexylamines as contact insecticides elaborated the importance of minor changes in basic toxic moieties upon optimum toxicity to insect pests, safety to plants, and upon the convenience and economy of manufacture.

Mode of Action Studies.

These have been the primary preoccupation of insecticide toxicologists since the advent of the arsenicals. Early and influential investigations focused upon the degeneration of cells of the midgut following arsenical ingestion, vacuolization of nerve tissue exposed to the pyrethrins, and upon the effects of nicotine on the insect heart beat and of rotenone upon insect respiration. Progress was hampered by lack of basic knowledge of insect enzymology and neurology.

The advent of DDT in 1939 and the rapid proliferation of cyclodiene and organophosphate insecticides coincided with an exploding knowledge of fundamental biochemistry. The organophosphate esters (OP's) from wartime research were immediately recognized as cholinergic in action resulting from the inhibition of the synaptic membrane-bound acetylcholinesterase. This specific biochemical lesion was shown to occur not only in the target insect pest but also to pose major toxicological hazards to man and higher animals. Important discoveries made by insecticide toxicologists were the *in vivo* conversion of phosphorthionates (P=S) to phosphates (P=O) thus involving the delay factor in poisoning the role of the high energy phosphate bond in the inhibition of acetylcholinesterase, and the quantification of enzyme inhibition and lethality (LD_{50}) with the electron-withdrawing properties of aryl-substituents in the parathion/paraoxon type of insecticide. Exhaustive comparisons of insect acetylcholinesterase with the comparable mammalian enzyme showed appreciable differences in the active site of the enzyme that could be exploited in the development of selective OP insecticides. The high selectivity of malathion and dimethoate to insects over mammals was cogently explained in terms of the comparative pharmacodynamics of detoxication. Studies of the fate and transport of OP's in plant tissues led to the development of a sound methodology for use of systemic insecticides.

From these insights into the mode of action of OP insecticides, insecticide toxicologists became leaders in the study of the generalized effects of OP's on non-target organisms. Studies of the delayed neurotoxic syndrome in vertebrates including man, demonstrated that the well known "ginger jake" paralysis produced by tri-*o*-cresyl phosphate was due to the *in vivo* formation of a cyclic saligenin phosphate that reacted preferentially with a key "neurotoxic esterase" that promotes the growth and regeneration of long axons of the nervous system. Several groups of OP insecticides, *e.g.*, the alkyl and phenylphosphonates and their thioates were demonstrated to cause irreversible degeneration of the nervous system and consequent irreversible paralysis. These compounds were shown to inhibit neurotoxic esterase following chronic dermal or oral exposure or after single dose exposures at very low levels. These studies were important in elucidating the causal agents of mass episodes of human delayed neurotoxicity and resulted in the withdrawal of a number of very hazardous commercial insecticides.

Mode of action studies with acetyl cholinesterase inhibitors led to the development of the aryl *N*-methylcarbamates as enzyme inhibitors and insecticides that are synthetic neurohormone analogues. These carbamates differ from the OP insecticides in that the carbamates have a high affinity for the active site of acetylcholinesterase together with a relatively slow turnover rate of the carbamylated enzyme.

Mode of action studies with DDT were largely inconclusive and led to many divergent views. These have been largely reconciled by electrophysiological investigations that demonstrated a prolonged after potential following depolarization of the nerve axon, resulting from a delay in the closing of the Na+ channels. This action is a function of Ca++ dependent protein kinases. The pyrethroids act upon the same basic ion-gating mechanism of the axon. The action of the cyclodienes was also shown to involve the nerve axon and the lesion was eventually defined as the blockage of *gamma*-aminobutyrate synaptic transmission by competitive inhibition of the action of this neurotransmitter.

Insect growth regulators modeled after the structures of the insect juvenile hormones were devised by insecticide toxicologists. Hundreds of structurally optimized analogue mimics have been shown to interfere with the development of physiologically competent adult insects and thus arrest the development of the insect in the juvenile stages. An entirely new type of insecticidal compound, the chitin-synthesis inhibitors were discovered serendipitously but have been shown to interfere with the complex biochemistry of cuticle formation and thus to produce a critical biochemical lesion at molting, pupation, or ecdysis.

Structural optimization of the pyrethrins, the components of the naturally occurring pyrethrum insecticide from *Chrysanthemum cinerariaefolium* has been exceptionally rewarding in the development of more effective insecticides. Understanding of the molecular centers that are attacked during biodegradation of the pyrethrins has led to the development of synthetic pyrethroids that have broad spectrum activity and lengthy persistence. These insecticides are an order of magnitude more effective than other conventional types and during the 1980s have become major components of insect control programs.

Insect Resistance to Insecticides.

Resistance to the action of insecticides is a major challenge to economic entomology and was first recorded with the San Jose scale in 1914. Within two years hydrogen cyanide resistance was observed in California red scale and in black scale. By 1940 resistance to lead arsenate had been demonstrated in the codling moth and in the peach twig borer and to tartar emetic in the citrus thrips and the gladiolus thrips. The discovery of DDT in 1939 and the very widespread use of this insecticide during the 1950s when yearly production in the United States reached a maximum of 163 million pounds, led to the rapid onset of resistance in the house fly, the human body louse, and the bedbug. Cases of DDT resistance rose from 3 in 1948 to 13 in 1954 to 98 in 1970 and to 229 in 1980. These fascinating examples of "accelerated microevolution" inaugurated intensive study of resistance phenomena that have remained a major preoccupation of insecticide toxicologists for nearly 50 years. Fundamental to the characterization and measurement of resistance were the technics of topical application and log-probit analysis of dosage-mortality curves. Radioisotope methodology, a product of World War II research, became indispensable and was coupled with *in vivo* and *in vitro* studies of insect biochemistry. Major progress was made in characterizing one type of DDT resistance as the result of enhanced *in vivo* detoxication of DDT to its non-insecticidal ethylene, DDE, a reaction catalyzed by a unique enzyme, DDT'ase, which is regulated by a semidominant gene on chromosome 11 and requires glutathione as a co-factor. A variety of structural analogues of DDT were shown to be DDT'ase inhibitors, *e.g.*, chlorfenethol, bis-*p*(r-chlorophenyl)-chloromethane and 4-chloro-*N,N*-dibutylbenzenesulfonanilide. These had a brief period of success as synergists to restore the activity of DDT against resistant insects. "Resistance proof" DDT analogues that could not be dehydrochlorinated by DDT'ase were also developed, *e.g.*, 1,1-bis-(*p* chlorophenyl)-2-nitropropane. These elegant strategies eventually came to naught as Ddt'ase levels became even higher and as another form of DDT resistance, *kdr* or target site insensitivity of the sodium channels in the insect nerve axon, became widespread.

As other types of insecticides, lindane and the cyclodienes, organophosphates, carbamates, and pyrethroids were introduced successively for insect control; insecticide toxicologists diligently studied their developing resistance mechanisms. Cydlodiene resistance was shown to relate to altered nerve axon receptors of the GABA-nergic transmitter, organophosphate resistance to enhanced aliesterase production, and carbamate and pyrethroid resistance to enhanced mixed function oxidase production. A more complex resistance mechanism, that of altered acetylcholinesterase was discovered, where this essential synaptic enzyme was shown to have changed its fundamental structure so that it no longer was inhibited by organophosphates and carbamates, Behavioral resistance was demonstrated in mosquitoes and horn flies. At present, the diversity of resistance mechanisms and their genetic controls have resulted in widespread multiple resistance in hundreds of insect pest species that no longer respond to available types of insecticides, *e.g.* the house fly, horn fly, German cockroach, malaria mosquitoes, the Colorado potato beetle, the diamondback moth, the chrysanthemum leaf miner, and the citrus thrips. Toxicological solutions are not promising and the mere measurement and characterization of insecticide resistance has lost its charm.

Environmental Toxicology

For more than a century, insecticide toxicologists have been concerned about the environmental fate of insecticides and their effects on non-target organisms. The first major concern was revealed in studies of deciduous fruit orchards in the Pacific Northwest where the heavy use of lead arsenate has resulted in accumulations of lead and arsenic in soils that prevented the successful replanting of young trees. Environmental toxicology really became a discrete discipline during the 1950/1960 period when residues of DDT and the cyclodienes were shown to have soil persistence measured in years and to lead to the bioaccumulation of metabolites such as DDE and dieldrin through food chains. Bioaccumulation of DDT/DDE residues in coho salmon and lake trout from Lake Michigan was found to span a million fold range from parts per trillion in lake water to parts per million in fish. Raptorial birds feeding on these fish were shown to suffer from impaired reproduction through the action of DDE in preventing proper egg-shell development. Ubiquitous residues of DDT/DDE, dieldrin, and oxychlordane were found everywhere in human fat and milk, and the cyclodienes in particular were demonstrated to produce abnormal central nervous system damage not only in fish and birds but also in humans where epileptiform convulsions were found to occur after heavy exposure during manufacturing and spraying.

These findings clearly defined the need for suitable pesticide biodegradability and insecticide toxicologists quantitatively explored the biodegradability of insecticides with laboratory model ecosystems using radiolabeled pesticides and in larger field scale plots. These studies appropriately defined the biodegradability of xenobiotics and made the term a household word. A series of biodegradable DDT analogues were produced using the model ecosystem technology to demonstrate this basic environmental attribute. These sorts of insecticide toxicological investigations by the 1980s were fundamental to a dramatically changed viewpoint about the nature of insecticide use and the importance of integrated pest management (IPM) incorporating safer and non-polluting insecticides.

General References

Metcalf, R. L. 1955. Organic insecticides; their chemistry and mode of action. Interscience, N.Y

Metcalf, R. L. 1990. Insect control, Ullman's encyclopedia of industrial chemistry, English edition.

O'Brien, R. D. 1967. Insecticides: action and metabolism. Academic Press, N.Y.

Sheperd, H. H. 1951. The chemistry and action of insecticides. McGraw-Hill, N.Y.

Wilkinson, C. F., ed. 1976. Insecticide biochemistry and physiology. Plenum, N.Y.

WHO 1971. Bulletin World Health Organization 44: (1,2,3), 1-470.

Insects as Vectors of Plant Pathogens

Norman W. Frazier

It has been estimated that insects appeared on earth 350 million years ago and humans 250 thousand years ago, and we have probably been battling the insects for control of the world in a losing war ever since. Since antiquity insects have spread diseases to humans, animals, and plants. They have destroyed our food and caused annoyances by biting, stinging, itching, and using us as a source of food.

Plant diseases that were insect-borne were known long before it was discovered that they could be spread by insects, *i.e.*, peach yellows was discovered in 1791 in the U.S.A. and potato-leaf-roll in Germany in 1845 and color break in tulip in the Netherlands was known even before then but was not recognized to be a disease.

It seems a timely coincidence that the first demonstration that an insect could be important as a vector of a plant disease occurred a century ago when in 1891 fire blight of pears was shown to be spread by bees.

During the ensuing 99 years, new demonstrations of an insect vectoring a plant pathogen were relatively rare, until 1911 when the subject was first reviewed. Other reviews followed into the 1920s when the relations of insects as vectors of plant pathogens were realized to be a very important facet in human welfare, thus becoming an important and fascinating field for study and research, up to the present.

The relationships between the insects, the susceptible host plants, the plant pathogens and the environment are so complex and the literature on the subject voluminous in detail that space permits only a sketch of such relationships.

Four primary factors are involved in the transmission of plant pathogens by insects, viz: the host plant, the insect vector, the disease agent, and the environment. Also involved are the nature of the agents, their interactions with the vector and the host plant, *i.e.*, whether the agents are transmitted by the vector in a nonpersistent, semipersistent, persistent, or internal manner; in a passive manner externally by contact as in pollen transmission by bees and other insects; whether the agent multiplies within the insect, or can be transovarially passed through the egg. An additional basic relationship is the specific plant tissue inoculated, whether epidermal, mesophyll, phloem, or xylem.

A list of plant pathogens transmitted by insects and others includes the following:

Bacterial - over 200 species causing blights, rot, leaf spots,wilts,hyperplasia (galls) and a serious disease of grapes and alfalfa (Pierces' Disease) transmitted by leaf hoppers and spittlebugs.

Fungal - over 66 species of fungi causing 45 diseases spread by over 100 species of insects were estimated to occur in 1958. Fungi can also be disseminated by wind, water, man, and other animals.

Toxins - very common on numerous plants by the feeding of toxicogenic insects, primarily of the order Hemiptera and Homoptera, including leafhoppers, aphids, psyllids, mealybugs, plant hoppers, plant bugs, etc., all of which have sucking mouth parts.

Virus - nucleoproteins or molecules, possibly the most commonly vectored plant pathogen, transmitted by virtually all of the groups of insects that vector plant pathogens, as well as by mites, nematodes, and humans.

MLO - (mycoplasmalike organisms) transmitted mainly by leafhoppers, fulgorids, spittlebugs, and perhaps by aphids also.

XLO - (xylem-limited bacteria) transmitted by leafhoppers and psyllids.

Spiroplasma - transmitted by leafhoppers.

Nematodes - vectored by weevils, long-horn beetles and flies.

A provisional list of insect vectors, the number of transmitting species by group, the number of diseases transmitted and the kind of pathogen transmitted is contained in Table 1. Nematodes are included in the table because, although taxonomically they are not insects, most people think of them as such, and they are vectors of some very important diseases. Fungus and toxin incited diseases are not included since many fungal diseases are not dependent on insects for their spread and toxins are not pathogens in the traditional sense (recovery follows depopulation of the toxicogenic insect).

There are over 200 known plant diseases including some very damaging ones for which no vector is known.

The war between humans and insects obviously has not yet been won by us. In fact, we may be losing it because of our efforts to battle with toxics that poison our atmosphere, soil, water, environment, and ourselves. A better strategy would be the use of clean culture, biological control, pest management (Integrated Pest Management) and the continued development of molecular biology toward the creation of pest resistant or disease immune crop plants.

Most of this text and table has been based on information gleaned from the publications cited in the following list.

Table 1. A provisional list of insect vector groups, the number of transmitting species, and number of diseases transmitted, and the kind of pathogen transmitted.

Vector	Number of Species	Diseases	Pathogen Transmitted
Aphid	192	167	Virus, MLO
Leafhopper	130	76	Virus, MLO, RLO, spiroplasma, bacterium
Whitefly	3	22	Virus
Mealy bug	19	4	Virus
Plant bug	2	1	Virus
Thrips	4	2	Virus
Grasshoppers	2	5	Virus
Beetle	7	6	Virus
Treehopper	1	1	Virus
Fulgorid	11	7	Virus, MLO
Psyllid	4	2	MLO, RLO
Flies	1	1	Virus
Spittlebug	10	2	Virus, MLO, bacterium
Nematode	22	17	Virus

Suggested Further Reading

Carter, W. 1962. Insects in Relation to Plant Disease. John Wiley & Sons, Inc. 705 pp.

Evans, H. E. 1984. Insect Biology. Addison-Wesley Publishing Co.,Inc., 436 pp.

Harris, K. (ed.). 1984. Current Topics in Vector Research, Vol. 11., Praeger Publishers, New York.

Harris, K.F. (ed.). 1988. Advances in Disease Vector Research. Springer-Verlag, Inc., New York, 300 pp.

Harris, K. F., & K. Maramorosch. (eds.) 1977. Aphids as Virus Vectors. Academic Press, Inc., New York, 559 pp.

Harris, K. & K. Maramorosch. (eds.) 1982. Pathogens, Vectors and Plant Diseases: Approaches to Control. Academic Press, 310 pp.

Leach, J. G. 1940. Insect Transmission of Plant Diseases. McGraw-Hill Publications in the Agricultural Sciences, 615 pp.

Maramorosch, K. & K. F. Harris. (eds.). 1979. Leafhopper Vectors and Plant Disease Agents. Academic Press, 654 pp.

Maramrosch, K. & K. F. Harris. (eds.). 1981. Plant Diseases and Vectors, Ecology and Epidemiology. Academic Press, New York, 368 pp.

Metcalf, C. L., W. P. Flint, & R. L. Metcalf. 1951. Destructive and Useful Insects. 3rd ed. McGraw-Hill Book Co, Inc., New York,1070 pp.

Smith, K. M. 1951. Recent advances in the study of plant viruses. The Blakistan Co., Philadelphia, 2nd ed. 300 pp.

Smith, R. F., T. E. Mittler, & C. N. Smith. (eds.). 1973. History of Entomology. Annual Review Inc., Palo Alto, 517 pp.

Veech, J- H. & D. W. Dickson. 1987. Vistas on Nematology: a Commemoration of the Twenty-fifth Anniversary of the Society of Nematologists. Society of Nematologists, Inc., Hyattsville, Maryland. 509 pp.

Insect Behavior And Naturally Occurring Chemicals

Donald A. Nordlund

The insect world is awash in chemicals. Not manmade insecticides, but chemicals produced and released by insects, other animals, and plants. Many of these chemicals play important roles in interactions between organisms, ranging from mating and oviposition to feeding and defense.

Chemicals that influence interactions between organisms are called semiochemicals. The influence of semiochemicals on insect behavior is a rapidly growing field of study, in part because it is so interesting, and in part because it may be possible to use these naturally occurring chemicals in more biorational pest management programs.

Semiochemicals that influence interactions between organisms of the same species are pheromones. The existence of pheromones has been known since the early 1600s, when Charles Butler noted that bees are attracted and provoked to mass sting by a chemical released by a single sting. Serious study of pheromones, however, did not begin until the late 1950s. There are sex, alarm, trail following, aggregation, and epideictic pheromones, to name a few.

Many insects use pheromones to attract and seduce members of the opposite sex. In many moth species the females attract males from considerable distances using pheromones released from a gland at the tip of their abdomen. When the male arrives he will release a pheromone from glands on his wings, that makes her receptive to mating. Initially it was believed that the female produced attractant consisted of one compound. However, we have found that they often consist of several compounds that occur in very specific ratios. There is a great deal of interest in sex pheromones as a means of monitoring and controlling pest insects.

If you have ever disturbed an ant mound or termite nest and watched them respond, you have seen how an alarm pheromone works. Many insects that live in large nests or that tend to congregate produce alarm pheromones which might produce either a fight response, as in the ants, or a flight response, as in aphids.

Ants also provide the best known examples of trail following pheromones. They are able to maintain the trails that are so familiar to us by depositing a pheromone on the substrate. Should you remove some of the pheromone, for example, by moving a leaf that they are crawling over, the ants will become confused and run in all directions. Eastern tent caterpillars also follow trail pheromones from their nest to prime feeding locations and back.

Many species of cockroaches tend to aggregate, at least during periods of rest, in response to aggregation pheromones. Southern pine beetle females release frontalin when initiating an attack on a pine tree. Initial boring results in the release of α-pinene from the pine tree. The combination of these two chemicals attracts other beetles to the tree, resulting in a massive attack and the death of the tree.

Epideictic pheromones are important in regulating population density. Larvae of the Mediterranean flour moth secrete an epideictic pheromone, from glands at their mandibles that tends to regulate density by encouraging emigration of larvae from crowded areas, lengthening the time required for larval development, reducing the fecundity of resulting adults, and by inhibiting oviposition by adult females. Epideictic pheromones often have very subtle effects. Parasitoids also use epideictic pheromones to discriminate between healthy hosts and those which have already been parasitized.

Allelochemicals are involved in interactions between organisms of different species. One can easily visualize three types of chemically mediated interactions, between organisms of different species: one in which the receiver benefits, one in which the releaser benefits, and one in which both benefit. Thus, there are three types of allelochemicals: kairomones, allomones, and synomones, respectively.

How do mosquitoes find you so fast when you are out on a warm summer night? You produce a kairomone, lactic acid, which attracts them. Predator-prey or parasite-host types of interactions often involve kairomones. For example, *Trichogramma pretiosum*, a minute wasp that is parasitic on insect eggs, is stimulated to search for host eggs by kairomones in moth scales, to drill into the egg by kairomones in the material used to attach the egg to plant surfaces, and to deposit an egg by kairomones found inside of the host eggs. In this case we have a chain of behaviors with each link influenced by a different set of chemical stimuli.

Allomones, on the other hand, are often used in defense. The stink bug is deserving of it's name. If you ever have the opportunity to catch one you will find that they release a rather odoriferous substance that will remain on your hands for sometime. The bombardier beetles are able to spray an antagonist with a 100°C solution of quinones. Most plants possess chemicals that make them unpalatable to many insects that otherwise would feed on them. Venoms used to subdue prey are also allomones.

A bug in Java (*Ptilocerus ochraceus*) feeds on a particular species of ant, which is attracted by an allomone produced in a gland on the underside of the bug's body. When an ant of the right species approaches, the bug raises up the front of it's body so that the gland is visible and accessible to the ant. Should an ant of the wrong species approach, the bug will not accept it and, in fact, presses the front of its body closer to the ground. An accepted ant will feed at the gland while the bug, without attempting to feed on the ant, holds onto the ant's head with its fore legs. After a while the allomone in the gland begins to paralyze the ant, and when it shows signs of paralysis, the bug seizes it firmly, punctures it with its proboscis, and sucks it dry.

Synomones are often involved in mutualistic interactions. Various floral scents, for example, attract insects that feed on the nectar and pollinate the flower in the process. The parasite *Trichogramma pretiosum* is also attracted to tomato plants where it might find suitable host and thus, the plant benefits by the increased mortality of a herbivore and the *Trichogramma* benefits by an increased probability of finding hosts.

Semiochemicals play some extremely important roles in insect behavior. The complex web of semiochemical mediated interactions that exists in nature provides the entomologist with many interesting research problems. These naturally occurring chemicals may also contribute to improvements in our ability to control insect pests.

Suggested Further Reading

Barbosa, P. and D. K. Letourneau (eds.). 1988. Novel Aspects of Insect-Plant Interactions. Wiley, New York. 362pp.

Bell, W. J. and R. T. Card (eds.). 1984. Chemical Ecology of Insects. Sinauer Associates, Sunderland, Massachusetts. 524pp.

Nordlund, D. A., R. L. Jones, and W. J. Lewis (eds.). 1981. Semiochemicals, Their Role in Pest Control. Wiley, New York. 306pp.

Biotechnology Developments In Entomology

George J. Tompkins

Biotechnology can be defined as the manipulation of living organisms and/or biological processes to provide useful products. Within the scope of this definition the first applications of biotechnology occurred thousands of years ago when people discovered that microorganisms could be used in fermentation processes to make bread, brew alcoholic beverages, and produce cheese. Although biotechnology is not new, many of its tools such as recombinant DNA, hybridoma technology, immobilized cells and subcellular components (enzymes), bioreactors, and electrophoretic separation techniques are relatively new tools.

Recombinant DNA technology has and will continue to benefit many areas of entomology including developmental, behavioral and evolutionary biology, ecology, and applied entomology. The combination of molecular biology techniques and computer analysis should greatly aid insect systematics and population genetics. This should help in clearing up some of the problems in species identification within some disease vector complexes.

A gene is a sequence of nucleotides in a DNA molecule that specifies a particular function. Gene expression is a process of gene activation and the genetic information is expressed resulting in a specific product such as a protein. Knowing where a gene is located on the DNA molecule, and how it can be activated or repressed, is integral to the science of biotechnology. Restriction enzymes are a class of enzymes acting specifically on a DNA molecule by cleaving or cutting the DNA at specific nucleotide sequences. The purpose of these enzymes is to defend the cell against foreign DNA by cutting and inactivating the foreign genes. Hundreds of restriction enzymes have been catalogued, listing the specific nucleotide sequence they will cut in a DNA molecule.

Recombinant technology or genetic engineering is basically the incorporation of a gene from one organism into the genetic material of another organism. This technology can be used to enhance the ability of an organism to produce a particular chemical product or to enable an organism to produce an entirely new product. The basic steps in a genetically engineered recombination are: (1) identify the gene that directs the production of the desired substance, (2) isolate the gene using restriction enzymes, and place it into a separate piece of DNA, (3) then transfer the recombinant DNA strands into bacteria or other suitable host organisms, and (4) to clone many copies of these host bacteria that contain and express this new gene.

Plasmids are a circular form of DNA found in the cytoplasm of bacteria and plasmids replicate in the cell separately from the chromosome. A gene can be spliced into a plasmid with the help of another class of enzymes called ligases, which seal together the ends of the DNA segments. The resultant recombined DNA is then inserted into the host organism. The plasmid acts as a vector, a vehicle for transporting DNA into another cell. Once inside the bacterium, the plasmids replicate producing exact copies of themselves. Each clone is an identical copy of the recombined DNA and faithfully reproduces the gene for the production of the desired protein. In the early period of development the most commonly used alternate host was the bacterium *Escherichia coli*; however, now a variety of other hosts including yeasts, algae, and the cells of higher animals and plants can be used to express and study foreign protein.

Molecular biology may be of great help in defining many of the factors determining the abundance or distribution of insects. Molecular biology can benefit entomologists in determining how insects behave and respond to environmental factors such as food, temperature, and mates. In the last two decades studies on insect control by endocrine manipulation utilizing juvenile hormones and ecdysteroids has progressed rapidly. These hormones are simple molecules that regulate the processes of molting and metamorphosis. The disruption of these processes could effectively control pest insects and, at present, several juvenile hormone analogs are commercially available for control of flies, mosquitoes, cockroaches, and fleas based on disruptions of metamorphosis and reproduction.

The neurohormones are another category of insect hormones which are generally peptidic and have greater structural diversity than most of the ecdysteroids and juvenile hormones. The insect neurohormones, although only a few have at present been isolated and chemically defined, may have greater potential for insect control than do the ecdysteroids and juvenile hormones. The neurohormones in insects regulate many physiological processes related to reproduction and general physiological homeostasis, and also regulate the secretion of juvenile hormones and ecdysteroids.

One of the areas where molecular biology and genetic engineering will have a tremendous impact on applied entomology and pest management programs will be in the use of microbial pesticides. There has been an aggressive effort in many countries to engineer more effective viruses, bacteria, and fungi for pest control. These studies are broadly aimed at altering host specificity, increasing virulence, or broadening the range of environmental conditions in which a pathogen can survive and infect its host. Studies on epizootiology are necessary to understand and take advantage of natural mortality caused by microbials.

The major species of bacteria which can be used to kill insects are spore-forming bacilli of which there are two major groups: *Bacillus thuringiensis* and *Bacillus sphaericus*. A variety of strains exist within each group, and these strains differ in structure of toxins, their mode of action, and their targets. Some strains of these microorganisms are toxic to mosquitoes which are vectors of major diseases worldwide. The delta-endotoxins of Lepidoptera specific strains of *B. thuringiensis* are currently the most intensely utilized microbial insecticides, and several transgenic plants containing the *B. thuringiensis* endotoxins have been produced. Other strains of *B. thuringiensis* that have toxicity to early instar larvae of the Colorado potato beetle have also been isolated. A great concern is that the extensive and intensive use of *B. thuringiensis*, especially if continuous treatment and exposure of pest populations is achieved through toxin production in transgenic crop plants, may produce widespread resistance in currently susceptible pests.

The use of biotechnology to produce potato plants that have an insect repellent in the leaves is another example of economic insect control. A gene for leptine, a chemical that repels insects, was recently placed in potato plants by cell fusion. The Colorado potato beetle has developed resistance to most chemical insecticides in recent years and this biotechnology discovery may be of significant value in the control of this important potato pest.

Many diseases spread by insect vectors occur throughout the world. Projects directed towards the development of vaccines including malaria, dengue, anthrax, and others are presently being undertaken. Immunologists believe that it may be possible to inoculate humans against the bite of mosquitoes. This would not protect the person from the disease, but would kill the insect vector. This is similar to a study in which a vaccine injected into cattle resulted in the death of the ticks that bit them, as the cattle developed certain antibodies that interfered with the tick's digestive system.

In controlling insect and weed pests of economically important plants, many chemical pesticides are used. Biotechnology may offer some solution to the problem of safe and inexpensive disposal of aqueous pesticide wastes generated by pesticide applicators or farmers. The use of genetically engineered microorganisms more efficient in partially or totally destroying pesticides in wastewater solutions would be greatly preferred to the current practice of storing the waste material in open pits and allowing naturally occurring organisms and sunlight to degrade the pesticide components.

The desire of the scientific community to ensure socially responsible research and development with recombinant DNA research led in 1975 to an international conference held at the Asilomar Conference Center in Pacific Grove, California. This conference was sponsored by the National Academy of Sciences with the support of the National Institutes of Health(NIH) and the National Science Federation (NSF) and led to the creation of the NIH guidelines for recombinant DNA research. This process has led to the current regulatory procedures under the directorship of the Environmental Protection Agency (EPA), the Food and Drug Administration (FDA), and the U.S. Department of Agriculture (USDA).

The views expressed are those of the author and do not necessarily represent those of the Agency.

Chapter 6

Protecting Our Environment and the Challenge of Entomology as a Career

THE FAR SIDE   By GARY LARSON

How entomologists pass away

Protecting the Environment

John W. Kliewer

The "Environmental Movement" has deep roots. There are several notable historical events which have called attention to the importance of preserving and protecting our environment. The path has been tortuous and controversial but has lead to where concern for the environment is now certainly one of the major issues of our time.

Of interest in the environmental movement are the actions of individuals that have influenced the course of events. The trail has been blazed by such notable figures as John Muir whose efforts to save wilderness areas resulted in the creation of both Yosemite and Sequoia National Parks about one hundred years ago. Rachel Carson's 1962 book, "Silent Spring," has more directly affected entomology because it called attention to the possible devastating effects that the misuse of pesticides could have on the environment.

April 22, 1990 was the 20th anniversary of "Earth Day". The year 1970 has been called "The Year of The Environment." That year saw not only the first Earth Day but also a number of other environmental landmarks such as the birth of the Environmental Protection Agency (EPA), the enactment of the National Environmental Policy Act, the creation of the President's Council on Environmental Quality, and the passage of a new Clean Air Act.

Now the 1990s have been declared "The Decade of the Environment." The President has indicated he would like to be known as the "Environmental President" and, in keeping with that theme, has taken steps to elevate the position of Administrator of the EPA to Cabinet rank. And the momentum continues to build.

Entomology, too, has recently celebrated an anniversary. In 1989 the Entomological Society of America commemorated the one hundredth year of "Entomology Serving Society" so it is fitting to review just how entomologists fit into this environmental picture.

Many of the critical environmental issues of our day involve the science of entomology. Some of these problems such as pesticide contamination of food, water, air, and soil, are closely related to our profession. These problems involve both rural and urban situations and occur both outdoor and indoor.

Entomologists are most directly associated with those environmental issues having to do with the control of insects, mites, ticks, and other pests. Researching and monitoring effectiveness of pesticides, recommending appropriate control measures, and developing alternative methods to control pests which destroy crops, spread disease organisms, devour stored products, damage homes is the domain of entomologists.

Environmental concerns were not overriding issues in the early days of pest control. However, members of our profession have become increasingly aware of these problems. Integrated Pest Management (IPM), which relies on various control means which reduce the need for total reliance on chemicals to control pests, has ascended to a dominant position. Entomologists have played important roles in initial development of these environmentally preferred IPM methods and it is clear that responsibility to protect the environment is even more important today.

Entomology, laws, and regulations

The use of pesticides and the laws which regulate pesticides in the environment are high profile entomology-impacting topics of conversation these days. This public concern about exposure to and the risks (or perception of risks) associated with pesticide use should not be under-estimated.

Entomologists have figured prominently in the evolution of the laws and regulations which govern today's use of pesticides. Enforcement of such regulations falls primarily to EPA and related state agencies.

The first federal law concerning insecticides was passed in 1910 to protect farmers and other users of these chemicals. Before that time, many useless preparations were advertised and sold. Even those which were effective, for example, Paris green, were often adulterated and, in the 1860s, several people spoke out against such unethical practices. Benjamin Walsh, State Entomologist for Illinois, was especially outspoken in his crusade against these practices. The Federal Insecticide Act of 1910 served well and put an end to nearly all of the fraud in the sale of insecticides.

Prior to World War II, pesticide legislation was intended simply to protect the user from adulterated products. However, after WWII new chemical pesticides were developed that were often more toxic than the early pesticides and they were recognized as being capable of causing harm to nontarget species when improperly used. As a result, the first Federal Insecticide, Fungicide, and Rodenticide Act — FIFRA — was enacted in 1947. This Act required that all pesticide products be registered and that all pesticide labels list the contents of the product. For the first time, pesticides were required to be registered before they could be marketed in the U.S.A. Among other things, the 1947 FIFRA required that "economic poisons" carry labels bearing certain information, including warnings intended to prevent injury to humans. An economic poison which failed to comply with the labeling requirement or which could not be rendered safe by any labeling, was "misbranded." The Secretary of Agriculture, who had the responsibility to administer the Act until EPA was created in 1970, was required in that case to refuse or cancel the registration of the economic poison. No other enforcement remedy was provided to the Secretary. Now, under the amended FIFRA, there are a number of options available to EPA such as the restricted use classification.

Congress has amended FIFRA several times since it was enacted in 1947. Those of 1972 were especially important. Referred to as FEPCA (the Federal Environmental Pesticide Control Act), they shifted the emphasis from safeguarding the consumer against fraudulent pesticide products and their possible lack of efficacy to a pesticide's potential for causing harmful effects to nontarget species, human health, and the environment.

Recently, on October 25, 1988, the President signed additional major amendments to FIFRA. Among other things, these amendments required a substantial acceleration of the reregistration process for previously registered pesticides and authorized the collection of fees to support reregistration activities. The law also changed EPA's responsibility and funding requirements for the storage and disposal of suspended and canceled pesticides and the indemnification of holders of remaining stocks of such canceled pesticides.

Environmentally sound entomology

As mentioned, entomologists generally are strong advocates of IPM procedures and have figured prominently in helping to bring about the changes necessary for less dependence on chemical controls in pest management. Some of the alternative control measures that have been developed are: (1) biocontrol measures, that is, use of pathogens, parasites, and predators, and (2) use of naturally occurring chemicals such as pheromones and insect growth regulators (IGRs), which are generally more compatible with the environment than are most pesticides.

The use of bacterial agents which are pathogenic to certain pests is currently expanding. A strain of *Bacillus thuringiensis* (*israelensis*) is successfully used for control of mosquito and blackfly larvae. Another strain (*kurstaki*) is effective against caterpillars and two other strains (*tenebrionis* and *san diego*) have been recently registered by EPA for control of the Colorado potato beetle. More research is being done with this pathogen and it is likely that it will find other target pests.

Examples of predators in pest control range from insects and mites that are predacious on pest species to the use of fish such as *Gambusia* and other minnows for mosquito control. One particularly interesting application of biological control involves the use of a group of mosquitoes, *Toxorhynchites*, which transmit no diseases and are not even a nuisance to man or animals. The larval stage of these mosquitoes prey on the larvae of other mosquitoes some of which may be serious pests of man and animals. Control measures of this type are attractive because the predacious species can naturally invade environmental niches (tree holes, for example) which can be exceedingly difficult if not impossible to find and treat.

Pheromones (sex attractants) find their greatest utility in monitoring for the presence of pests but they can also play roles in disrupting mating activity and capture. While IGRs, which prevent maturation of target species, have been registered and are effective, they, as standard pesticides, have the potential to adversely affect non-target species as well. With the current and increasing concern for endangered species, IGRs are to be used with care and in accordance with label instructions. Well planned IPM programs take advantage of such measures as those mentioned above but also are generally based on careful scouting in order that pesticides, when needed, are used judiciously and with minimum threat to the environment.

The importance of such environmentally sound alternative pest control measures can scarcely be overemphasized and entomologists have generally been in the forefront of this changing concept.

Possibilities in the field of alternative pest control are exciting and virtually unlimited. Recent developments in biotechnology (genetically engineered pathogens) for example are fascinating. There are, in fact, a number of exciting avenues of research, and while one might think that there is little left to be discovered, this supposition is not true. A saying attributed to a wise old Indian from the mountains of western Montana or some equally pristine place might apply here. Legend has it that while discoursing on the subject of acquired knowledge, he noted that "the larger the island of the known, the longer the shoreline of the unknown". There is still much to discover in the fascinating field of Entomology.

The challenge is for environmental compatibility.

Suggested Further Reading

Carson, R. L. 1962. Silent Spring. Houghton Mifflin, Boston. (25th Anniversary Edition, 1987. Houghton Mifflin, Boston. 448 pp.)

Mittler, T. E., F. J. Radovsky and V. H. Resh. (Current Editors). Annual Review of Entomology. Vols. 1-36.

Price, L., J. Heritage, K. Flagstad, J. Lewis, R. Barker, and M. Rogers (Current Editors). EPA Journal. Vols. 1-16. Issues of particular interest for this article: (1) Regulating Pesticides, Vol. 10(5), June,

1984, (2) Agriculture and the Environment, Vol. 14(3), April 1988
and (3) Earth Day, Vol. 16 (1), Jan./Feb. 1990

When The Rocks Beckoned — Biocontrol Revisited

John J. Drea

It looked like an ideal set up at the time. The field was faintly reminiscent of a New England field but it was on a hillside far above a small village in central Afghanistan. Instead of a rock wall around the field, as many farmers construct to clear the area for farming, here the rocks were arranged in long low piles scattered throughout the field. Nevertheless, these rock heaps looked ideal for collecting beetles and other wild life that may hide under them. So Clif, George, and I, provided with a cyanide jar and forceps, each went off in search of rare and showy specimens. Collecting was good but difficult considering the size of the rocks and the 6000 foot altitude.

The three of us had met a couple of weeks earlier in September, 1956, in the Afghan capital of Kabul to survey for and collect natural enemies of halogeton, *Halogeton glomeratus*, a toxic range weed that had been introduced into North America many years earlier. The weed had invaded millions of acres in the western states and was implicated in the deaths of many of the native and domesticated grazing animals. It was economically prohibitive to apply herbicides over so vast an area. Eventually, the concept of controlling the weed with natural enemies was considered, especially after the then recent success with insects imported to control St. John's wort, or Klamath weed, *Hypericum perforatum*, in the western United States.

We were returning to Kabul after a two week survey in the Bamian Valley, with its fantastic statues of standing Buddahs, and in the neighboring valleys of the Hindu Kush, a chain of the Himalayas reaching into Afghanistan. It was lunch time so we had the driver stop in one of the rare shady areas next to the field. The rock piles beckoned and we answered their call. However, their's was not the only call we heard. Far away at the bottom of the hill an Afghan was waving and yelling at us. So we waved back. Pleasant people! Back to turning rocks and collecting beetles. The calls got louder and louder as the man approached. He was a formidable looking individual, complete with baggy pants, a long knee-length shirt, a rather not so clean turban, but a bright and shiny rifle, and a bandoleer of bullets across his chest. The long curved knife at his waist and the turned-up shoes on his feet almost completed the picture. The dark and angry look on his face did complete the picture. Since I was the closest, I stood up, smiled, and added "Salaam Alaikum", the extent of my linguistic abilities at that moment.

The Afghan looked at us, slowly shook his head when I showed him my collection of beetles, and then began to gesticulate. He stood tall as only a man with a loaded rifle can, looked up at the skies, put his two hands together as if he was holding an open book, and then put a hand behind each ear like he was listening for something. This was certainly confusing coming from someone who was so angry. Gradually, the light began to dawn. I had seen these same gestures by others in the streets and fields of Kabul and when passing through Karachi to Afghanistan. They were part of the ritual associated with the call to prayer that devout Muslims respond to five times a day. Why did this man yell at us, run up the hill fully armed, and then, in a mood obviously not conducive to prayer, start to pray? Were we the objects of a "conversion or sword" tactic practiced by too many zealots over the years? He kept looking up at the sky and going through the prayer ritual. Suddenly, he put his two hands together beside his head in the universal pose of a sleeping person. That is when it hit me! These were not rocks piled up for farming. These were gravestones and we were tearing them apart. We were desecrating the local cemetery.

I turned and called to Clif and George to stop what they were doing, carefully put the rocks back and get over here to apologize to a somewhat upset local inhabitant. We made a *big* show of putting the rocks back in place, requesting his approval with each move. When he grasped that we were not ghouls or grave robbers or any other similar forms of low life, the tension eased. When the task was done, we all smiled and made strange noises to each other. Although offered cigarettes, all he took was the matches. I am grateful to this day that he was an understanding individual and not quick on the trigger. When we returned to the van, Mohammed, the driver, was still asleep with no idea how close he had come to preparing a strange trip report.

After Afghanistan the halogeton project was moved to Tehran, Iran, and in 1959 to Rabat, Morocco. Unfortunately, we never did find an acceptable biocontrol agent and the United States Department of Agriculture closed the project down in the spring of 1963. However, to paraphrase Murphy's Law, if things can go wrong they will wait until the last minute to do so. Murphy struck again!

A week before ending operations and moving on to other projects in other countries, we had received a large bottle of potassium cyanide, a very deadly powder used for killing insects, among other things, which had been ordered from headquarters many moons before. We did not need it now nor did we know what to do with it.

Packers and movers were everywhere at the laboratory. Therefore, I brought the bottle home but, because of our small children, I put the bottle on a top shelf in the garage, pending proper disposal of the chemical.

When I returned from the laboratory the next day, a day before we were to leave the country, the garage had been cleaned out of everything, including the wooden shelves and the cyanide. My wife had told the maid that she could have whatever was in the garage, including the shelves since wood was scarce and expensive. She did not know about the cyanide. The maid's husband had come with a donkey cart and removed everything. Unfortunately, we did not know where the maid lived, except that she was from the Arab quarter of the city. Furthermore, her name was Fatima, as was just about every other female inhabitant of Morocco, in deference to Fatima, the daughter of The Prophet. Almost every male used the name Mohammed, or a variance of it for a similar reason. To call after someone in the street you just yelled "Fatima" or "Mohammed" and you got everyone's attention. Then, you picked out the one you wished.

Somewhere in the Arab quarter Fatima and Mohammed had possession of a pound of cyanide and neither of them could read the label or anything else in English. At that time, most of the Arab population did not have family names as we know them. Identity was based often on lineage, such as Mohammed ben (son of) Mohammed ben Ali ben Abdul, or Fatima bent (daughter of) Mohammed ben Rashid ben Moustapha. Therefore, we could not look for Mrs. "somebody". What was the next step? We decided to go into the Arab quarter that evening and look for a Fatima who just had a new baby, had worked for an American, and was out of a job. The meager description fit a large number of the population.

We went to the residential area and began asking questions, not to locate who had taken our cyanide (we would have been given a chase that would have made the proverbial goose appear tame), but on the pretext that we owed Fatima money due her. Within minutes, we had a following with offers of all sorts of names and houses. Eventually a well informed individual told us that the Fatima we were after was at a wedding in Meknes (a city about 150 km. away) and would not be back until late that night.

We returned around 11 p.m. and re-gathered our crowd. The Fatima we met was charming and insisted we come in for tea. However, she was three times the size of our Fatima. It looked hopeless and we were ready to go to the police and explain the problem. By then, the crowd around us was huge but sympathetic and as helpful as possible.

A little lad in this vast field of faces said that his mother used to work for an American but her name was not Fatima. What did we have to lose? So we followed the little fellow to his house. When the door of the house opened I don't know who was the most surprised, my wife and I or Fatna (her name really wasn't Fatima after all). We explained the problem, thanked the crowd for their help, and went into the small apartment. Immediately, we asked for the key to their storage bin and her husband, Mohammed, and I searched the storage area. The bottle of cyanide was there, deadly but unopened.

The rest of the night was spent in celebration, tea drinking and general good company. However, one statement Fatna made really startled me. She was amazed that we were so concerned about the welfare of the Arabs in the quarter that we would dare to come there in the middle of the night to look for her. I don't think she had considered the alternative. I doubt that my wife and I will ever forget our last night in Morocco and the charming people we met in the Arab quarter of Rabat. Incidently, the cyanide was turned over to the American Embassy and placed in a locked safe. It may still be there.

From Rabat, we were transferred to the USDA's European Parasite Laboratory (EPL) in the outskirts of Paris. On a cold snowy day in March, 1963, we arrived in France. I had been informed that I would be assigned to look for natural enemies of the face fly, *Musca autumnalis* DeGeer, a serious introduced pest of cattle and other animals in North America. Unfortunately, this involved research into the dung habitat of the fly. Consequently, I was designated Chief Chip Checker with instructions to "check 'em out."

A dairyman I was not. Milk came in cans and bottles and I was happy with the arrangement. Nevertheless, duty called and I armed myself with a set of plastic trays, rubbermaid gloves, and a very long handled serving spoon, thereafter known as the "pooper scooper". I never did get used to sitting on the ground, surrounded by chips in all shapes and forms, as well as the producers of this storehouse of biological diversity, and sifting through this fragrant medium. It just wasn't my bag.

The face fly maggots are easy to see in a chip. They are bright yellow, and when they are there, they are there in numbers. No problem. Actually, the cows are the problem. They are extremely inquisitive and if they come up behind you just to see what you are doing — they are capable of very quiet movements — they can nudge you right into the medium.

My wife and I developed a technique that worked quite well when she and the children accompanied me on a collecting trip. The family would gather at a far end of the field, outside of the fence, and begin to dance, sing, and wave their arms. These antics would attract the attention of the cows, and anyone passing. The animals would invariably walk, trot, or even gallop over to watch what we referred to as the "Bovine Ballet". All too often so would the passersby. These antics would give me a chance to get into the field, check the chips, collect the promising deposits, and get out before one of the herd would see me and spread the word that interesting things were going on in another part of the pasture.

Chip checking in other parts of the world brought us many strange looks and reactions. While we were collecting in Greece one herder watched us moving from cow pad to cow pad, finding some of immense interest while rejecting others of apparently the same appeal, and then, of all things, scooping out the contents with a soup spoon. He stood it as long as he could. He came to us and asked "Deutsche?" (German). We said no. We were Americans. He looked at us for a very long minute, muttered "Americans!", put his forefinger to the side of his head and wiggled it in the universal sign indicating insanity, and marched off without another word. I wonder what he told the villagers.

During the same trip, but in Turkey, we came upon a very productive cow pad area. Unfortunately, it was at the outer wall of a little (thank heavens!) village. We stopped, got out our pooper scoopers and then clandestinely tried to make off with the harvest from the cows. But nothing in Asia passes unnoticed. Within minutes, or even seconds, we had a bunch of children all around us presumably trying to decipher why we were in the chips. At first, we tried to remain serious but scooping poop does not lead to developing a learned and dignified impression. Therefore, we decided to have some fun in the dung. We literally skipped to each pad, flipped the dried crust off the top and stirred up the contents to catch sight of any yellow larvae within. If there were some, one of us would let out a cry of delight and we'd both scoop like mad. The kids could not resist this game. Soon they, too, became immersed in the game (they didn't have scoopers) and would scramble from chip to chip. When called, one of us would examine their find and either make a long face while slowly shaking his head or let out a cry of joy and gather at the chip to happily scoop the contents into our plastic bowl.

Frankly, with the arid conditions of the Middle East, we found few flies but we left a large number of impressed children. In some cultures, insane people are considered to have been touched by God and therefore are special. I often wondered how often we were considered "special" and, consequently, were left alone during our wanderings in out of the way places.

Summer projects at the EPL included the collection of the migratory locust, *Locusta migratoria* L., and its associated parasites. The parasites obtained were destined for eventual release in grass-hopper-infested area of the western U.S.A. Although plagues of locusts usually conjure up biblical scenes in the Middle East and Africa, over the years, locusts (grasshoppers) have devastated great areas of the United States. Control of these beasts was difficult and even impossible. Maybe the flies that lived on the hoppers in Europe could be used to combat these pests in North America.

A major difficulty in working with the migratory locust is that it can fly fast and far. Yet we needed large numbers of the insect to recover any parasites that may be living in this flying machine. Techniques and, surprisingly, advice were rare commodities when it came to collecting the host insect. Collectors had tried aerial nets but could not get close to the shy hoppers. Some even folded long strips of heavy plastic to swat the little beasts. Whatever was devised was not very efficient. We needed a new technique.

Hoola hoops were in fashion at the time. They were light and fairly rugged to meet the demands of the little dears who swiveled in them. When fitted with a loose netting the hoops had promise as a means to trap the grasshoppers. With some practice and considerable effort, the hoop worked well.

The hoppers fed and bred in Les Landes, a beautiful region of sand dunes, pine forests, fields, streams, and delightful country restaurants in southwestern France. Unfortunately, the hoppers preferred less inviting environments. They lived in logged-over areas that were cluttered with slash and tree stumps and clumps of dense grass growing between the young saplings. Consequently, many things besides grasshoppers lived in these fields.

The collecting was difficult. The name of the game was to tramp through the field and scare up a hopper, follow its flight until it landed, and while keeping an eye on the spot where the hopper had settled, run to within about 10 feet of this landing site, look for movement, and sail the hoop as hard and fast as possible at the place the hopper was presumed to be. Then came the fun part. The collector ran to the hoop, went down on hands and knees and tried to catch the hopper while it was struggling to escape. However, other things came up in the net. The area was a favorite haunt of a local venomous viper and a choice site for various types of yellow jackets to nest. As a result, the collector was never really certain what would be captured in the net. The collector, not the grasshopper, was often the first thing away from the hoop. Nevertheless, we were able to have several hundred hoppers on a good day, and several close calls on a bad day.

We were trying to collect several species of parasitic flies of the genus *Blaesoxipha*. These flies were interesting in that they ovaposited on the hopper while the host was in flight. They would rest on the ground waiting for the hopper to jump. When the host flew the fly would attack and parasitize the insect.

My wife, Leyla, accompanied me on several of the trips as a collector and cage carrier. On one trip, before we had recognized this "wait and fly" response of the fly, my wife was following after me with a bucket covered with a net in which we placed captured game. I was having a terrible time finding the host. Although it was a hot day, the hoppers would not cooperate, the ground was rough, and I was covered with bruises from tripping over branches and falling into holes. My wife was having her troubles, too. She was constantly bothered by dozens of pesky flies that would land on her and either rest or stroll all over her. They would not let her alone. We found out later these were the *Blaesoxipha* we were after. She was doing all she could to chase away these miserable beasts and I was doing all I could to capture them. Had I put a net over my wife and gathered the flies on her I would have saved us both hours of collecting and grief.

The hoop technique was a great success. However, we did have several motorists drive off the road into the fields while trying to figure out what these adults were doing with hoola hoops in the middle of a field among all the pine trees.

One of the more pleasant projects that competed favorably with the face fly project was the survey and collection of natural enemies of the cereal leaf beetle, *Oulemma melanopus* (L.). This exotic chrysomelid was established in the Great Lakes area and was a serious threat to small grains. In the early 1960s the EPL was ordered to look for parasites and predators of the beetle. In addition to collections in France I went to Italy to see what I could find.

The first step was to locate infested fields and collect and dissect numbers of the beetle larvae to determine if they were parasitized and the degree of parasitism. Then collecting was concentrated in those fields with the most parasites.

The collecting technique was simple. The collector cut off the portion of the leaf that bore the beetle larva or egg. This bit of leaf was transferred to a large cage with a double bottom. When it had finished feeding, the mature larvae dropped through the screen to pupate in the substrate of the second bottom. Under the right conditions of humidity, temperature, and other factors, eventually the adult beetles or the parasites emerged. The major disadvantage of this technique was that almost all of the leaves of the plants in a well infested field were removed along with the insects.

With an Italian assistant, Peter, we had surveyed much of Italy. Early in the season we had found a most promising field in the outskirts of Rome. The field was about a third of a football field in size, and was well infested with hosts heavily parasitized. We just had to collect in this field! Presumably the owner lived in the small house next to the field. So, we knocked on the door and explained our present purpose in life. We asked him if he objected to our working in the adjacent field. His response was that he thought it was a good idea and he didn't mind. Off we went and for the next four to five weeks we wreaked havoc in the field. Not deliberately but it is next to impossible to tramp through a wheat field on a daily basis without "cracking eggs" so to speak. We waved to all those passing by and they waved back. We were secure, with the blessings of the owner.

The field did pay off. We recovered several species of parasitic wasps including the tiny egg parasite, *Anaphes flavipes* (Foerster), that now helps control the cereal leaf beetle in the U.S.A. By midsummer collecting was finished and I closed up shop to return to Paris. As a parting gesture, I went to the U.S. Embassy and bought a bottle of hard stuff (expensive as blazes on the local market). Peter and I drove out to the farmer's house and knocked on the door for the second time that season. When he appeared we were all smiles and explained how successful we had been. We thanked him profusely for letting us use his field. His response was, "It's not my field." He did not get the booze. Peter did.

Over the years the entomological experiences and the personal adventures arising from these foreign assignments were exciting and rewarding but I would be hard pressed to decide which were the most memorable. It was an excellent period of time to be in the "foreign service." Although political upheavals and terrorism were going on somewhere all the time, they seemed, at least to me, to be more directed toward particular goals and less randomly targeted as they are in today's world. Foreign exploration is still an exciting experience. But, I am glad I was there when I was — when the rocks beckoned!

Suggested Further Reading

Coppel, H.C. and J.W. Mertins. 1977. Biological Insect Pest Suppression. Advanced Series in Agricultural Sciences: 4. Springer-Verlag New York, Inc., 175 Fifth Avenue, New York, N.Y. 10010, 314 pp.

DeBach, P. (ed.). 1964. Biological Control of Insect Pests and Weeds. Reinhold Publ. Corp., New York, N.Y., 844 pp.

Graham, Frank, Jr. 1985. The Dragon Hunters. Truman Talley Book - E.P. Dutton, Inc. 2 Park Avenue, New York, N.Y. 10016. 334 pp.

Papavizas, G. C. (ed.). 1981. Biological Control in Crop Production. Beltsville Symposium in Agricultural Research: 5. Allanheld, Osmum & Co. Publ. Inc., 81 Adams Drive, Totowa, N.J. 07512, 461 pp. 1981.

van den Bosch, R., Messenger, P. S. and A. P. Gutierrez. 1982. An Introduction to Biological Control. Plenum Press Publ. Corp., 233 Spring Street, New York, N.Y. 10013, 247 pp.

Microbial Control of the Japanese Beetle

Samson R. Dutky

The introduction of the Japanese beetle into the U.S.A.

In 1916 the Japanese beetle was discovered on plants in a nursery in Riverton, N.J. by Harry B. Weiss. It became established before there were restrictions on the introduction of plants into the U.S.A. The Japanese beetle spread rapidly since there were no natural enemies to control it. The Japanese beetle lab was established in Moorestown, N.J. to look into this problem. The insecticide used then was lead arsenate at rates of 500-1500 lbs./acre of turf, but that method of control did not keep up with the spread of the beetle. Japanese beetle traps containing floral extracts were reasonably effective but they also brought in thousands of other beetles endangering all of the plants in the immediate area. Introduced parasites of the Japanese beetle from Japan and Korea were quite effective but didn't have the capacity to spread with the beetle. Nematodes such as *Neoaplectana glaseri* were somewhat effective. Dr. Glaser developed methods for growing nematodes in artificial media but nematodes were still not as effective as milky spore disease.

The discovery of the milky disease microorganism

Under the direction of Dr. Selman A. Waksman of Rutgers University, the discoverer of streptomycin, I was chosen in 1933 to investigate the possibility that a microorganism could be weapon of control for the pest since it appeared that the beetle population was diminishing in the area of introduction. I was assigned to the Japanese Beetle Laboratory at Moorestown, N.J. where we worked cooperatively with the N.J. Agricultural Experiment Station. Within the first several weeks alert field workers brought in unusual and sick Japanese beetle grubs from which we isolated a spore forming bacteria identified as *Bacillus popilliae*, the milky spore disease. This organism multiplied in the hemolymph of an infected beetle grub until they became infected with millions of spores. Methods were developed to propagate the microorganism in larvae to the spore stage and a microinjector was used to inject approximately 2 billion spores/ grub. Later techniques were developed to dry spores and prepare a powder which retained its infectivity.

The testing of the milky spore disease organism

Initial tests indicated that introduction of the microorganism by feeding was possible but not profitable. Living diseased larvae were collected. Less than two months after the first recovery of *B. popilliae*, injected larvae were placed in the field to check for the spread of the disease and how long it took an area to be inoculated. Workers noted heavy populations of Japanese beetles which were free of disease in Bridgeton, N.J., so tests with *B. popilliae* by "planting" injected larvae verified the ability of *B. popilliae* to establish and spread in the Japanese beetle grub population. Tests were conducted on the effect of "planting" at different times of the year. A spore dust was developed that could be applied to the soil or turf in a grid pattern. Ralph White measured the spread of *B. popilliae* on Staten Island, N.Y. as starlings fed on milky diseased Japanese beetle grubs and spread the microorganism in their droppings.

Eventually communities and large areas were treated. For example, the people of Doylestown, Pennsylvania, didn't want to use chemicals and chose to treat all the turf with milky disease. The great success of this program helped a lot in publicizing this as a preferred control method. All the park areas in Washington, D.C., including the mall, were treated and protected against the Japanese beetle. During World War II all the turf at airports was treated with milky spore powder. The state of Maryland treated one acre in every square mile with milky disease. We provided inoculum and instructed their workers. A very important lesson resulted from an experience when I was in the service in World War II. Until that time we had control of the inoculum and checked regularly by testing that the microorganism we propagated was indeed *B. popilliae*. While in the service less experienced workers failed to perform the proper tests, the yield/insect dropped and the organism they began to propagate was not *B. popilliae*. Fortunately this was finally corrected.

The Japanese beetle is still spreading. It took years to cross the mountains in Pennsylvania but it has probably been spread by people, *e.g.*, campers, in their laundry as they bundle it up and travel from day to day.

In March of 1945 we had a sudden warming, then the temperature dropped. The temperature in the warm spell was high enough to germinate the spores within the beetle grubs. Then the temperature dropped. At a soil temperature of 50 F. the vegetative rods in the insects were shadow rods and were destroyed by the insects' defenses and when an extended period of cold came, the rods were killed and the spores did not produce infected insects. So the first brood was not controlled but the next brood was controlled. Fortunately the insects were starting to decrease to lower levels by the time that I was to receive an award at the fall meeting of the association.

Three patents that I obtained were one on the method of controlling Japanese beetles, a method of propagating the bacteria, and a microinjector for inoculating the beetle grubs.

Choosing a career in research.

Looking back, my research was fascinating and I was very pleased at making such an important contribution as I was able to work with many people in many areas and had wonderful experiences over more than a forty year career.

Added note.

The Dutky method of production has been used by the Fairfax Biological Laboratory, Clinton Corners, New York for more than forty years to produce a commercial product called "Doom".

Further Reading

Bulla, L. R.,Jr., Costilow, R. N., & Sharpe, E. S., 1978. Biology of *Bacillus popilliae*, Advances in Applied Microbiology. 23:1-18.

Dunbar, D.M. and Andreadis, T.G. 1975. Present status of Japanese and oriental beetles in Connecticut, J. Econ. Entomol. 68:453-457.

Dutky, S. R., 1940. Two new spore-forming bacteria causing milky disease of Japanese beetle larvae. J. Agr. Res. 61: 57-68.

Dutky, S. R., 1947. Preliminary observations on the growth requirements of *Bacillus popilliae* Dutky and *Bacillus lentimorbus* Dutky. J. Bacteriol. 54: 257.

Dutky, S.R., 1963. The milky diseases, in Insect Pathology: An Advanced Treatise, Steinhaus, E.A.,Ed., Vol.2, p. 75-115, Academic Press, Inc., New York.

Fleming, W.E., 1961. Milky disease for control of Japanese beetle grubs, U.S.D.A. Leaflet No. 500, 6 pp..

Flemimg, W.E., 1970. The Japanese beetle in the United States, U.S.D.A. Agricultural Handbook No. 236, 30 pp..

Klein, M.G. 1988. Pest management of soil inhabiting insects with microorganisms in Agriculture, Ecosystems and Environment vol.24: 337-349, Elsevier Science Publishers B.V., Amsterdam.

Klein, M.G., Johnson, C.H. and Ladd, T.I., Jr.. 1976. A bibliograpghy of the milky disease bacteria (*Bacillus* sp.) associated with the Japanese beetle, *Popillia japonica* and closely related Scarabaeidae. Bull. Entomol. Soc. Amer. 22: 305-310.

Tashiro, H. 1987. Turfgrass Insects of the United States, Cornell University Press, Ithaca, New York, 391 pp.

U.S.D.A., 1970. Controlling the Japanese beetle, U.S.D.A. Home and Garden Bull. No. 159, 16pp..

Training an Entomologist

Marvin K. Harris

The study of insects has attracted many people because insects are biologically interesting, spectacularly beautiful, amazingly abundant, and exist side by side with man wherever humans inhabit the globe which ensures their ready availability as well. The modern entomologist certainly appreciates these insect features, but typically has the privilege of studying insects because of the threats some pose to human valued resources. One such insect is the pecan weevil, *Curculio caryae* (Horn), whose larvae attack the pecan kernels just before the nut can be harvested. This pest generated a great deal of concern among pecan growers in Texas and they prevailed upon Texas A & M to recruit an entomologist to address the problem.

They hired me and one of my first duties was to become familiar with the problem. My formative years on a Nebraska farm and at Dana College, and graduate work on apples and pears at Cornell helped some, but pecan trees and pecan weevils were new to me and I had a lot to learn quickly if the university and the growers were to be pleased.

As luck would have it, the grower research and extension committee met in Fort Worth a few weeks after I was hired and they asked me to attend. I knew they had pushed hard to establish my position and would probably expect some preliminary answers or at the very least a sound plan on how I would proceed to solve their problem. After all, I had a Ph.D. in such matters from Cornell and what was difficult if not impossible for the common man should be just a minor irritation for one with such an education. I didn't yet know pecan weevils (the little dickens has a 2-3 year life cycle and you can't learn all about it in 3 weeks, even if you resort to books), but I knew if pecan growers were anything like the farmers I grew up with, this was going to be an experience. They respected education but were intolerant of the pompous or arrogant and they prized common sense above all other forms of knowledge. And their favorite sport at the local pool hall, where I received my gallows education, was baiting the 'college boys' with questions like "why does my corn moulder in the crib?" delivered with a naive helplessly ignorant tone begging for some relief from this plague. The subject would try to ignore the question if they didn't know the answer, or braver ones would offer a few hypotheses that might impress some ivory tower professors. The crowd would thicken and quietly but gleefully the questioner would persist until either the educated victim was clearly shown not to know or had given a highfalutin answer like: "The moisture content of your corn was too high at harvest and this has allowed growth and development of fungi on your stored grain, hence the mouldering." "Now that's a good'un!" they'd say, "That there answer just beats all!" they'd cry, and a chortle would wave through the crowd as the questioner would turn to the town dummy who was always consulted as the final authority in these matters for the answer. "She was put up wet" he would reply in this case and usually add "You fool, you put 'er up too wet and now she's mouldering — serves you right!" Then the questioner would turn back to the college boy to say "Is that what you said?" Often as not the college boy was gone. I had a few reservations about meeting with the pecan growers so soon.

The session started quietly enough with discussion of routine business that involved pecan show results and orchard tours. The chairman presiding was a self made septuagenarian multimillionaire who ran department stores for a living and grew pecans out of pride, with profit an important but secondary issue. This was clear when the subject turned to my hiring and the pecan weevil problem. The friendly amiable demeanor of the chairman was maintained as he introduced me as Dr. and noted I'd completed my education at Cornell. Several at the table allowed how they would make an effort to speak more slowly and more clearly out of deference to that fact. I said thanks and noted that I expected to develop a research program over the coming months that would address the pecan weevil problem. They observed that they would be satisfied with useable answers and the latest high technology was not a prerequisite for their requirements. Nice perhaps, but not essential. Fair enough.

The snack break ended our discourse and I thought I'd made as good an impression on the group as a Ph.D. could ever hope to on first meeting. I stayed for a cola. The chairman reconvened the session and reached into a small sack to remove something that looked like the oval white slow release aspirin caplets. He rolled several toward me and two arrived within reach while the rest careened this way and that on the polished surface of the table. "What do you think of that?" he asked with an invective in his voice that would have made a break-in burglar tremble.

A closer inspection in my palm proved they were cold, frozen probably. I removed my hand lens. Every economic entomologist should carry a hand lens, a notebook, a pencil, and a sharp pocket knife (if it isn't sharp then refuse to loan it as a matter of principle because every dude wants to measure you by its edge). This must be, I thought, a frozen pecan weevil larva. The head capsule, spiracles, and segmentation were obvious upon magnification, but there are dozens of weevil species whose larvae look more or less alike. The group was obviously waiting for an answer or comment. "It sure looks dead to me" I said.

When the laughter died down the chairman passed the sack down to me as well. Inside was an off white fleshy frozen pouch apparently filled with caplets. I scraped on it a little with my knife and puzzled over it with my hand lens racking my brain over their version of why corn moulders in the crib knowing that their confidence could be established or shaken depending on the outcome.

When you truly don't know something, the best course of action is usually to own up to it rather than to bluff. "I'm here to work with you to help solve your pecan weevil problem." I said. "Sometimes I'm going to need your help and this is one of them for I don't know what this is." The chairman looked around the group and for the first time I realized they were puzzled too. No answer from that quarter.

"It's a crop," he said. "I raise turkeys," and it was clear. He had released turkeys into the pecan orchard when the larvae were exiting the nuts and this bird had filled its crop with them. "They really like those redheads (the grower term for weevil larvae because of the reddish head capsule)" he said, "but they still don't eat enough to control them."

"Well," I said, "I'm glad you decided to give that job to a different kind of turkey," and I was too. The chairman recently passed away but during the intervening years with his help and that of equally cooperative growers we figured out how to control pecan weevil well enough to deny his birds a good meal in the orchard. And the experience has been another education altogether.

Commercial Entomology for Fun and Profit

Eugene J. Gerberg

At a fortuitous meeting at Camp Lee, Virginia in late 1943, a seed was planted that germinated to become an active, growing firm of consulting entomologists. Steve Easter, at that time a civilian entomologist for the Third Service Command, and I met. I was a shining new 2nd Lieutenant, U.S. Army Sanitary Corps. My first assignment was Assistant Camp Medical Inspector, and my task was to rid the camp of an epidemic infestation of bedbugs. Frankly, the only bedbug I had seen before this was on a microscope slide in a med ent class at Cornell. A call to American Cyanamid resulted in one of their representatives coming to Camp Lee for a quick teaching lesson on how to fumigate a building with cyanide. As Camp Lee was a Quartermaster camp, they were using the old WW I method of steam sterilization of the bed frames and mattresses. They left all the webbing belts, knapsacks, etc.,on the floor of the barracks. Examination of the webbing equipment revealed massive infestations of bedbugs. This is when I decided we had better fumigate the barracks, with all webbing equipment as well as beds and mattresses left inside. Over 700 buildings were fumigated without a single injury, except to the bedbugs.

Back to the formation of the company. Steve mentioned that he had two friends who would be interested in forming a company that would commercialize entomology. We contacted Ralph Bunn, at that time a Lt. Colonel in the Sanitary Corps, and Duncan Longworth, a Major. As all three Army entomologists were shipped overseas shortly afterwards, Steve carried the ball to form a corporation called Insect Control & Research, Inc., or ICR for short. We all returned to the U.S. by Jan 1946. As I was the youngest of the four, it was decided that I should be the one to start, and we should start in Baltimore, Maryland. Steve would stay on as a civilian entomologist. Ralph, by now a Colonel would stay in the Army, and Duncan would go back to the mosquito commission. We agreed that if for any reason, one could not be active in the company, he would sell his stock back to the company. We each had put a few thousand dollars into the company. With these few dollars, the company started operation on February 1, 1946.

In order to quickly bring some income into the company, we started in by providing a pest control service. In the day-time I would go out and sell our service, and in the late afternoon and evening, I would service the account. Steve helped whenever he could. After calling on a number of "food plants", it became obvious that they were in need of a professional service, that could inspect the facilities for insect and rodent infestations or other unsanitary conditions, and advise and assist in the correction of these conditions. The Food and Drug Administration began to more actively enforce the law, which in turn made our services more useful.

Shortly after we got started, Steve decided that as he had not traveled much, he would accept a position as entomologist with FAO. Ralph decided to stay in the Army, and Duncan went into practicing law. By the end of 1946, I was in need of help and Ben Krafchick, a Cornell schoolmate and fellow Army entomologist, joined the company. ICR steadily acquired more clients, and personnel and we divided the company into a "food-plant sanitation division" and a "pest control and termite division". We began receiving requests for consulting services, and started a "consulting division". One of the first jobs, was to make a mosquito survey and set up a program for a county in New York State. Aerial spray companies requested the services of entomologists, and we serviced most of the local companies.

In 1948, I decided to complete the requirements for a Ph.D. and registered as a part-time student at the University of Maryland. I majored in entomology and minored in plant pathology. Dr. Ernest Cory and Dr. William Bickley were my major professors. I received my degree in 1954. My thesis was "A revision of the New World species of powder-post beetles belonging to the family Lyctidae" which was published in 1957 as U.S.D.A. Technical Bulletin No. 1157.

In 1952, the Ministry of Agriculture of Venezuela, invited me to assist in setting up aerial spraying programs for agricultural crops in Venezuela. An American pilot, of Spanish parentage accompanied me on the month long trip. He turned out to be an excellent language instructor in addition to a superb pilot. Besides seeing agricultural activities throughout the country, it was an exciting time, as most of the trip was conducted under a state of martial law, due to the assassination of the President. From 1952 on, trips were made once or twice a year to Latin America, as a consultant for either governmental agencies or private companies.

In 1960 we were awarded a government contract to do research on mass rearing of mosquitoes, and to produce millions of sexed mosquitoes per month. This required specially designed rearing facilities, and so the start of our present facilities began. The contract was renewed each year for 5 years. A number of scientific papers resulted from our studies.

The U.S. Department of Commerce called me in the fall of 1961 and asked if I would be willing to go on a Trade Mission to Nigeria, representing U.S. interests in agriculture and public health. Seven of us toured Nigeria for two months, meeting with public officials, business men and anyone who wished to talk to us, particularly if it concerned the possibility of a joint venture, or some other business enterprise.

In 1965, the U.S. Army Medical Research & Development Command awarded ICR a contract to screen anti-malarial compounds. The original contract required mass rearing of suitable vectors, maintaining suitable hosts for the plasmodium parasites, maintaining quantities of infected vectors, and developing techniques for a rapid mass screening program for anti-malarial drugs. It was expected that we would be screening 30 compounds per month. The contract lasted 4 1/2 years. Before the contract ended we were screening over 1000 compounds per week! We developed a mosquito screen, in which the mosquito was fed the experimental compound, infected with the plasmodium, and then dissected to determine the presence or absence of oocysts or sporozoites. The test was extremely sensitive and highly replicable. It did require a large staff, and we had approximately 50 people engaged in rearing, feeding, and dissecting mosquitoes. An interesting sidelight developed from the need of a parasitologist. Walter Reed Army Institute of Research, requested that we include a parasitologist on our staff. I located a parasitologist, and asked my wife, Jo Betty, to accompany us to WRAIR where the parasitologist was to be interviewed. At the interview, my wife and I sat next to the parasitologist. The interviewer evidently did not catch my wife's name, and included her in the questioning. As my wife had studied med ent and parasitology at Cornell, she answered the questions, thinking that the interviewer was being polite, and including her in the conversation. The interviewer then called me outside and said, "Hire the woman." I told him it was my wife, and he said it didn't matter, hire her. So my wife became the parasitologist of the program. She did a great job, by the way, though I sometimes had to wait for dinner to be prepared.

The malaria contract required more space so we added a considerable amount of lab and work space to our facility, including a second floor.

The Department of Commerce called again in January of 1968, for me to join an "exclusive" Trade and Development Mission to Pakistan. This Mission was to be chosen from former members of Missions. It seemed like 30 days of luncheons and dinners, never knowing when you would be the guest speaker. All in all it was an interesting and exciting trip. After Pakistan, I continued on to India, where I met Ed Smith, trying to keep malaria under control. From there to Thailand, Hong Kong, Japan, and home.

In late 1968, Jimmy Wright, Chief, Vector Biology and Control at WHO in Geneva asked if I would be interested in few months tour as Acting Project Leader at the East Africa Aedes Research Unit at Dar-es-Salaam, Tanzania. In June 1969, I took 2 months leave and left for a weeks briefing in Geneva, and on to Dar-es-Salaam, where Dr. Tony Brown of WHO met me. The professional staff consisted of 2 young entomologists, Dr. Keith Hartberg and Dr. Milan Trpis, who became good friends of mine. The tour was extremely interesting, as one of my duties was to take Keith and then Milan on safari to various parts of Tanzania to look for mosquitoes. On my return trip to Geneva, I had the opportunity to stop at Nairobi, and visit with Mrs. Ellinor C. C. Van Someren. We designed and published a pictorial key to the *Stegomyia* mosquitoes of East Africa. The next year I had the opportunity again to spend 2 months as Project Leader of the Unit in Dar-es-Salaam. During a trip to Amani, Tanzania where I met with Dr. Graham White and Dr. Jim Hudson. On my return at the end of the tour, I visited Dr. Angus McCrae in Uganda.

In October of 1971, I was invited to Geneva to attend a meeting on screening anti-malarial drugs. After the meeting, I flew to Nairobi, at Dr. George Craig's request, and was met by Dr. Walter Hauser-mann. We drove to Mombassa, and moved the Mosquito Biology Unit (MBU) lab to an old but very adequate mansion on the beach. I was told that it was called the Gerberg Hilton for a while. I had an opportunity to again visit the MBU in 1974 (coincidentally mbu in Swahili means mosquito). When the malaria contract ended, it was evident that we put too many of our eggs in one basket, namely government contracts. We then concentrated on developing contracts in the private sector, particularly for pesticide testing and pesticide registration.

In 1973, we were approached by a prestigious Japanese company, Sumitomo Chemical Co., Ltd. and placed on a retainer. I became their International Technical Advisor. This has been a mutually very satisfactory arrangement. We then began receiving retainer contracts with a number of domestic companies. As we had one of the few louse colonies in the world (it originally came from the USDA colony in Gainesville, Florida), we tested most of the pediculicidal products for the U.S. producers. We also supplied various laboratories with cockroaches, flies, fleas, and other insects for testing purposes. One of the more bizarre requests was for thousands of roaches for a "horror film." We almost had to empty out our colonies to fill that order.

A trip to Japan, continuing on around the world, was one of the highlights of 1974. Sumitomo invited me to visit the company headquarters in Osaka and their fantastic laboratories in Takarazuka. Dr. Ross Arnett accompanied me on the trip. After Japan, we stopped for a week at the Seychelles Islands, to collect mosquitoes. We collected all but one species reported from the islands, and wrote a pictorial key to the mosquito larvae. From the Seychelles we went on to Mombasa, Hamburg, London, and home.

In the summer of 1977, I was asked by the U.S. State Department, Agency of International Development (AID), to make the first environmental assessment of a malaria control program. It was decided to do the malaria program in Sri Lanka. I spent a month, along with Henry Willcox, my assistant, assessing the situation on that beautiful island. The next year, AID asked me to do an environmental assessment of the malaria program in Thailand. Jay Graham accompanied me on that trip. It was a fascinating experience, as we examined the malaria programs from Chiang Mai to Songkhla, and from Bangkok to Trat. We also did an environmental assessment of the dengue fever program.

On the domestic front, our testing contracts continued to increase. We were testing household insecticides for most of the producing companies. The increase of testing and registration contracts required additional personnel, and Dr. Robin G. Todd was added to the staff. I had known Robin when he worked for Dr. Marco Giglioli in the Cayman Islands. In 1983 we supervised the agricultural testing of fenpropathrin, and had test sites in 32 locations around the country.

Between contracts from governmental agencies, universities, and private companies, we have managed to visit and work in most of the countries of the world. We have been able to collect beetles, butterflies, friends, and photos. We hope we have been helpful to our various clients. We must have done something right, as we are still in business. We had lots of fun and made a profit.

City Toads and Country Bugs

Eric Grissell

Most of the entomologists I know were born that way! An accident of birth some might say. Like the statistical findings in those doctor commercials, bugs were probably among the first loves in the lives of nine out of ten entomologists-to-be. That odd tenth-person, however, came to the bug world from slightly off-center ... most likely as a simple-minded naturalist. These directions in life, I believe, are beyond our own personal control, and are regulated by some random, certainly nonparental-linked gene. At least this is the hypothesis I've developed based upon my own entomological experiences during the last 25 years.

My early brushes with nature took place in as unlikely a naturalist's habitat as can be imagined ... the heart of San Francisco. As a kid, the city provided lots of inspiration for kid-type things. It was the best sort of pre-electronic kid-hood that the '50s could provide ... hide and seek, roller skating, running, jumping, jungle-gyms, five and dimes, soda fountains, ten-cent movies (a double feature every Saturday morning), endless back alleys, and miles of "stuff" to look at in every direction. The city was a visual, aural, and tactile stimulus of overwhelming proportion, but as kids go it was probably about average. My upbringing was about average too, a laborer father, housewife mother, and three-room apartment in as diverse a neighborhood as might seem possible. I was your basic normal kid in a basic city (though more beautiful than most), and I loved it.

For those predisposed to see, nature abounded in one form or other just about anywhere. And "just about anywhere" is where most city kids can be found. Every fence post, outdoor stairwell, or alleyway had its quantity of spiders for example. A friend and I found the abundant ebony-black and crimson-red spiders especially acceptable as playthings until ordered to stop by parental authority. Black widows, we learned, were not suitable playmates. Well, okay, if not spiders then worms or salamanders, cabbage whites or skippers, scotch broom, nasturtiums, honey bees, snails, pigeons, or even toads.

I suppose my eventual entomological adult life might be attributed (or even a tribute) to toads. Not a pretty picture, you might say, nor the thing one discusses in polite conversation. But toads are what really set me over the edge ... off the narrowed pathway that leads to normal employment as a fireman, dentist, or cowboy. It was an accident, as is most of life, that I discovered them in the middle of a city. I found a pond one day while wandering lost in unknown territory. I suppose in today's times I would have been in mortal danger, but in those days kids didn't have much to fear. (The most dangerous life form known to the incipient naturalist was the neighborhood bully, but having an intelligence not much greater than a slug, he was easily outwitted.)

I made several trips to the pond over the period of a month or so, until in the end it dried up. Eventually it was bulldozed to make way for some important city-thing or other ... row houses, I suppose. I could never find the area after the bulldozers went through. But for a month I was fascinated by the mire and muck, the polliwogs, dragonflies, water, vegetation, and eventually the toads. Finding this tiny bit of nature in the "big city" was the single most important event in my city life. It also embedded a mildly schizophrenic notion within me that cities and nature (and even naturalists) can be compatible ... maybe even synergistic.

It was not long after finding the pond that my family moved from the relative comforts of the city to the suburbs of the neighboring county. Perhaps wilderness would be a better word than "suburbs" for this was practically a national park by comparison. Here I could wander endlessly in a totally natural environment that was as varied as any adult naturalist could handle. At ten years old my senses were beaten senseless by the complexity of nature. This, and a non-stop reading habit, had me convinced at any one time that I would grow up to be a herpetologist, or a paleontologist, or an anthropologist, or a botanist, or an entomologist. I knew I would be one of those "... ists," I just couldn't figure out which one!

What eventually focused my attention on insects was apparently just a contest of elimination. I would have had a difficult time as a herpetologist, I finally realized, because my mother hated snakes and wouldn't let me keep them. She didn't much care for lizards or alligators either (thus came to an end as brilliant a career as any in the herpetology field). In all my day trips to the surrounding countryside I never once found a fossil or lost civilization, and that greatly hampered my development as a paleontologist or anthropologist. Out of desperation I eventually gravitated to what was most abundant in my habitat, namely plants and bugs.

Of the two, insects fulfilled a more immediate need than did plants. After all, chasing butterflies was a lot more stimulating than chasing dandelions. Anyway, what did you do with a plant once you caught it? From what I read you had to squash it flat, whereupon it turned black. Of what use is a flat, black mass of vegetable matter to a kid? It didn't make much sense. But a butterfly or beetle looked pretty much the same whether alive or dead. The choice, too, rested largely on what I could sneak into the house: my mother didn't think much of bugs (whose does?), but I could hide a lot more bugs in my bedroom than I could snakes ... take my word on it!

Once a path is chosen, there is no guarantee that it will lead to anywhere in particular or that it will extend past the next hillock. There is no guarantee that it is even the correct path. That is for the walker to decide. That is what life is all about. In my pathway there were no obstacles (other than the normal ones of time, money, and minor set-backs), but there were many friends who helped guide me from hill to hill, over valleys that sometimes took that extra-special effort only a friend could give.

When I needed a helpful hand as a child, Don MacNeill (then of the California Academy of Sciences) answered the many child-like questions I had about simple techniques of entomology. In all likelihood he would not now even remember me. Many scientists are like that ... helpful, friendly to anyone who has a serious question, no matter how dumb. Ellsworth Hagen, my high school science teacher, always took an interest in any interested student. Teachers are funny that way. He helped me think about insects and even to experiment with them for local science fairs. He introduced me to his cousin, Ken Hagen, who was a real-life entomologist at the University of California (who, many years later, sat on one of my Ph.D. examination committees). He also urged me to apply for a high school summer student grant through the National Science Foundation. This earned me a non-paid job for two months working as an entomological technician at U. C. Berkeley. My boss, Harold Madsen, chose me, I was told later, because I received a "D" in trigonometry and so had he as a high school student. Scientists, it was turning out, were not only helpful and friendly, but not necessarily much smarter than I. There was Ken Hobbs who gave me an entire collection of insects to supplement my Ph.D. studies. He didn't ask anything in return. I still owe him one for that. My major professor, Dick Bohart and his wife, Margaret, are the last to be included in my litany of entomological acknowledgments. They are here for completeness. From them I learned the human side of "academia" and I received much more than I could every repay in words, though I once tried to sum it up (1983, R. M. Bohart: Recollections 1964-1973: Pan-pacific Entomologist 59:16-20). There were, of course, many others along the pathway: professors, technicians, secretaries, my fellow peers, and students, all of whom coalesce into a picture of camaraderie beyond words. Each is a part of my education as we are all parts of each other's separate educations.

My interest and experiences with insects began as a young child and continued against all the typical childhood obstacles (i.e., ridicule, scorn, derision, gibes, contempt, lampoonery, mirth, hysteria, guffaws, cachinnation, snickers, chortles, etc., etc.) until I am now well into adulthood (some might say hopelessly beyond it). It is safe to say that childhood interest is a special commodity that comes from somewhere as yet unspecified. My interests were certainly not a reflection of my parents' (nor anyone in my known family). But, importantly, my parents allowed me the courtesy of self-determination (in all non-snake matters) and were "career neutral" so that I could choose anything I wanted to be even if they had no idea what it was. That I chose to be an entomologist was certainly a surprise to them, and sometimes, when I think overly long and hard about it, it is just as much a surprise to me.

Chapter 7

Unusual Facts About Insects and Other Arthropods

THE FAR SIDE　　By GARY LARSON

"Gad, I hate walking through this place at night."

Ancient and Modern Illustrations in Entomology

Marius Locke, Harry Leung, and Michael Locke

In 1665 Robert Hooke presented his "Micrographia" to the Royal Society. This included prints that were among the first detailed descriptions of insects. He figured and discussed the sting of a bee, the wings and foot of a housefly, the head and eyes of a drone fly, the egg of a silkworm, a complete bluebottle fly, the larva and pupa of a mosquito generated by the "putrefaction of rain water", adult male and female gnats, a plume moth, an ant, a body louse, and a silver fish. He was our first and most famous entomological illustrator. His illustrations have a freshness and clarity not seen again until the advent of scanning electron microscopy, 300 years later. The reason is that light micrograph photographs have a narrow depth of field whereas Hooke's views were reconstructed by the synthesis of information in many planes of focus.

Hooke's most famous picture may be that of the flea reproduced here (Fig. 1). He says of the flea "...the microscope manifests the beauty of it ... all over adorn'd with a curiously polished suit of sable Armour, neatly jointed, and beset with sharp pinns like bright conical Steel-bodkins; the head is on either side beautified with a quick and round black eye." The scanning electron micrograph of a flea shows how accurate and detailed Hooke's reconstruction is.

The scanning electron microscope (SEM) has several properties making it suitable for entomological studies. Its useful range of magnifications matches the size of insects and their surface patterns, allowing us to see and switch easily in the 100,000 fold range from one centimeter down to one tenth of a micrometer.

Secondly, it visualizes surfaces with a great depth of field, showing objects with all their parts equally in focus. SEM pictures are also very easy to comprehend. The light and dark of the images mimics the light and shade that we are accustomed to interpret in our everyday experience.

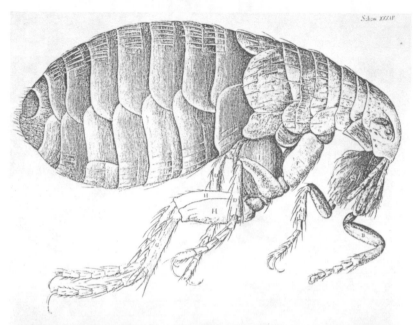

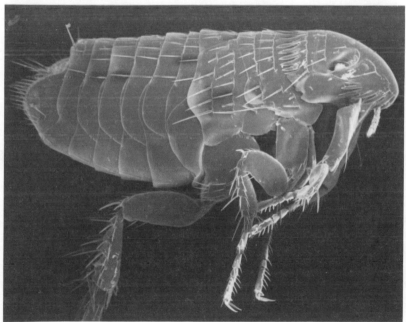

Figure 1.

Figure 2.

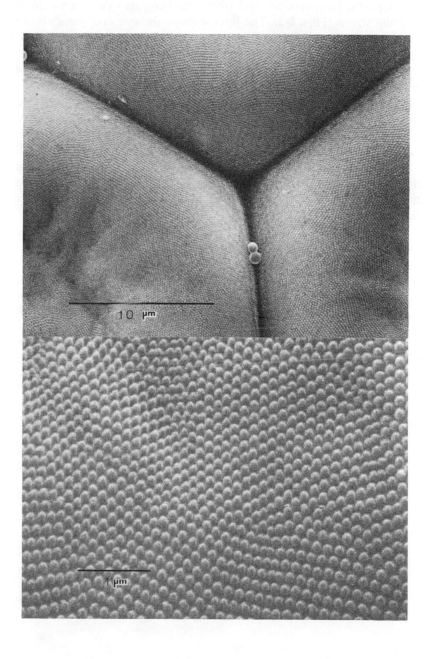

Figure 3.

We have illustrated the value of SEM for structural studies in entomology by taking the head of a butterfly (*Calpodes ethlius*, Lepidoptera, Hesperiidae) and photographing its eye at a range of magnifications (Figs. 2-3). The lowest show the eye and the scales in relation to the head. Intermediate magnifications resolve the individual ommatidia making up the eye. Higher magnifications bring out the arrangement of the surface cuticular pimples that function to prevent reflection like the blooms added to the surface of camera lenses.

Hooke concluded that "...Nature ... works by such excellent and most compendious, as well as stupendious contrivances ... that it were impossible ... to find any contrivance to do the same thing that could have more convenient properties". He had discovered how well adapted insect structures are to their function. Our modern scanning electron micrographs confirm Hooke's opinions. They reveal insects as beautiful, intensely complicated and superbly adapted to their ways of life.

Reference

Hooke, R. 1665, Micrographia: or some Physiological Descriptions of Minute Bodies made by Magnifying Glasses, with Observations and Inquiries thereupon. 270 pp., Jo. Martyn and Ja. Allefrey, Printers to the Royal Society, London.

Insects on Stamps

Denis R. Hamel[1]

Introduction

Insects are endlessly fascinating creatures, creatures most people think of as nuisances or pests. Unquestionably some are and we battle them incessantly. But though some are bothersome, they are no less interesting to those who will stop and study them.

Studies of the lives, habits, and structure of insects reveals their beauty, ingenuity, and ubiquity. Insects have always been a part of human life. In fact, this world we claim as "ours" is perhaps even more a world of insects. For millions of years insects have outnum bered all other forms of animal life. At present nearly a million species of insects have been described, but the final total may be twice or even three times this number. Not only are insects numerous in the number of species, they are also numerous in individuals — there may be 50,000 bees in a hive; 2,000,000 termites in a mound; or 40,000,000,000 locusts in a swarm covering hundreds of square miles, weighing upwards of 80,000 tons and eating their own weight in food every day! Insects have also become adapted to innumerable ways of life, in every conceivable habitat.

Compared with insects whose evolution began about 350 million years ago in late Devonian times, our human ancestors are relative newcomers, being traced back only about 3.5 million years. Obviously then, insects have been remarkably successful and deserve not only our attention but our admiration. And so they have by stamp designers from all parts of the world.

As of 1989, 289 stamp-issuing entities, from A (Afghanistan) to Z (Zimbabwe), have issued nearly 4,500 stamps with 1,817 different kinds of insects, or their close allies, on them. Of the 27 generally recognized insect orders, fourteen are represented on stamps.

The earliest postage stamps to show insect-related items depicted the beneficial aspects of insects as a minor part of the overall design. The first "insect" stamp depicted a small honeybee hive and was issued by Nicaragua in 1891 (Figure 1). Other countries soon followed suit and honored not only the beneficial honeybee but also silkworms, dragonflies, and sacred scarabs. It was not until 1939 that an insect pest was represented. It was then that the World Health Organization (WHO) sponsored its campaign to rid the world of

---

[1] Deceased.

malaria mosquitoes. In furtherance of the WHO campaign, Mexico issued its obligatory one cent postal tax stamp depicting the struggle of man and pest (Figure 2).

After World War II, the number of insects appearing on postage stamps increased dramatically. The first taxonomically identifiable butterfly on a postage stamp was a Raja Brooke's birdwing (Figure 3). It appeared in 1950 on a one cent regular issue of Sarawak. Although not in color on that first stamp, rainbow-hued postage depicting this butterfly and many others soon appeared from countries around the world.

Issues of stamps with insects on them became so numerous that checklists for collectors became highly desirable (see References).

Systematics

Insects form the class Insecta, by far the largest class in the phylum Arthropoda. They comprise the greatest number of all living creatures on earth, and although ancient, they probably have not yet climaxed evolutionarily.

Countless millions of insects have populated our planet since the Devonian. Even the number of species exceeds 70 percent of all creatures, and to calculate the number of individuals would be mind boggling. There are probably more than a million and a half species of insects and more await discovery. They are found in practically every part of the earth from the blazing tropics to the frozen arctic waters. Some 12 thousand extinct species have been described and more discoveries are yet to come.

During their evolutionary development, insects have adapted themselves so completely to so many environments and conditions that they have adopted some most improbable looking characteristics. Their diversity bears witness to their enormous vitality and biological adaptability.

There are so many interesting, attractive, useful, destructive, odd, and ugly insects in the world that it is no surprise that they have caught the imagination of many of the world's leading stamp designers. As philatelic treasures they can be collected with neither net nor notion that there will be any upsets in the balance of nature.

The following list gives a very brief description of each group of insects that appears on stamps from around the world:

1. Coleoptera is the largest order of insects and includes the beetles (Figure 4) and weevils (Figure 5). Their most distinctive feature is the structure of the wings. Beetles have four wings, with the front pair thickened into hard elytra that cover the membranous hind wings.

2. Collembola are minute insects that are also called springtails (Figure 6). They get this name from a forked appendage which most of them have on their abdomen that aids them in jumping.

3. Dermaptera or earwigs (Figure 7) are nocturnal insects that have a pair of harmless pincers on their abdomens. Earwigs are unusual insects because they brood their eggs and protect their young until they are old enough to fend for themselves.

4. Diptera are the mosquitoes (Figure 8, on back jacket) and true flies (Figure 9). They possess only one pair of wings (the front ones); the rear ones have been reduced to knob-like appendages called halteres.

5. Hemiptera or true bugs (Figure 10, on back jacket) are characterized by a half wing, *i.e.*, the basal portion of the front wing is thick and leathery, the apical half membranous. At rest the wings form a "V"-shaped pattern.

6. Homoptera is a varied order which includes aphids (Figure 11), plantboppers (Figure 12), and cicadas (Figure 13). They all suck plant juices.

7. Hymenoptera includes the social insects such as bees (Figure 14), hornets (Figure 15), wasps (Figure 16), and ants (Figure 17). They all, at some point in their life, have four membranous wings. Some of the most beneficial insects belong to this Order.

8. Isoptera means equal wings and refers to the termites (Figure 18), which when winged, have front and hind wings of the same size and shape. Termites are considered by many humans to be the most universally destructive insects.

9. Lepidoptera are the butterflies (Figure 19) and moths (Figure 20). The beautiful flying rainbows most of us have known since childhood. Their brightness is created mostly by colored scales on their wings. It is this characteristic that gives the order its name. Nearly two-thirds of all insect stamps depict lepidopterans.

10. Neuroptera includes the thread-winged neuropterids (Figure 21, on back jacket) and the nerve-winged owlflies (Figure 22, on back jacket). Insects in this order are considered the most consistently beneficial to humans.

11. Odonata includes the dragonflies (Figure 23) and damselflies (Figure 24, on back jacket). The name of the order is from the Greek word meaning "tooth." Both the nymphs and adults of dragonflies and damselflies are predatory on other insects.

12. Orthoptera contains many and varied insects, including grasshoppers (Figure 25), mantids (Figure 26), katydids (Figure 27), and crickets (Figure 28).

13. Ephemeroptera are among the most ancient of the winged insects. Their aquatic naiads are important fish food as are the

short-lived or "ephemeral" adults which are called mayflies (Figure 29).

14. Finally, the small order Mecoptera (scorpionflies and similar insects) are predators as both adults and larvae (Figure 30, on back jacket). These rather rare insects are harmless to humans.

Collecting representatives of these insect orders probably conjures up visions of a net, killing jar, and insect pins; however, such items are not needed. A collection of insects on stamps requires only pencil, paper, and a desire to have worldwide pen pals.

References

Hamel, D.R. 1991. Atlas of Insects on Stamps of the World, 500 pp.

Hank, A.J. 1984. Lepidoptera, Butterflies, and Moths. *Topical Times* 35(4):75-83.

Kramer, K. (In Press). Checklist Entomophilately. Netherlands, 400 pp.

Butterfly and Moth Stamp Society Checklist. 1986. The Swallowtail. London, England. 74 pp.

Smith, F.G.A.M. 1978. Insects on Stamps. Tring, Hertfordshire, England.

Stanley, W.F. 1979. Insects and Other Invertebrates of the World on Stamps. ATA Handbook No. 98. 140 pp.

Nishida, T. 1986. Butterflies on Stamps. Tokyo,

The present article is based on portions of a book entitled "Atlas of Insects on Stamps of the World" that Mr. Hamel has published. We are grateful to the Entomological Society of America for the loan of the negatives of the insect stamps used here.

Figure 1.

Figure 2.

Figure 3.

Figure 4.

Figure 5.

Figure 6.

Figure 7.

Figure 9.

Figure 11.

Figure 12.

Figure 13.

Figure 14.

Figure 15.

Figure 16.

Figure 17.

Figure 18.

Figure 19.

Figure 20.

Figure 23.

Figure 25.

Figure 26.

Figure 27.

Figure 28.

Figure 29.

Bible References to Insects and Other Arthropods

W. G. Bruce

Forward

Perhaps no other single book in the Western World is so frequently cited as the Bible. It is used to point a moral, to adorn a tale, to win an argument, and to title a book. This all-time best seller is also a rich source of quotable material on insects and their relatives — their records of devastation; their use by philosophers to exemplify desirable attributes of industry and humility; their ravages pointed to as retribution for evil-doing.

Before the advent of modern science, the Bible was often used as a bridge of antiquity. Although science is now able to antedate scriptural material, the Bible nevertheless remains a favored source. How could a dry, factual report of an excavation below the walls of Jericho compete with a colorful Near-Eastern poem, proverb, or history, translated into the robust English of Shakespeare's time! For this reason, the quotations given here are from the King James version. The sweep and dignity of its language make up for any inaccuracies modern scholars may have found in its translation.

For ease of reference, the author has listed the books of the Bible alphabetically rather than in the order of their occurrence. Most of us have forgotten their order, if we ever knew it. He has not included references to manna, which some entomologists believe came from a scale insect; this is debatable, so long after the fact. He has also left out references to honey, even though it comes from the bee; it is as divorced from the insect, once removed from the hive, as milk is from a cow when the milking is done. There has to be a point of separation somewhere.

Some may find the verse-by-verse references scanty, away from their context. This can be easily remedied by looking up the original, from the references given.

Any such compilation must have its shortcomings, but the author would not have made this compilation had he not believed that it would be helpful to someone.

Index of References

Insects and Relatives	No. of References
Ants	2
Bees	4
Beetles	1
Cankerworms	4
Caterpillars	9
Fleas	2
Flies	9
Gnats	1
Grasshoppers	10
Hornets	3
Lice	4
Locusts	24
Moths	11
Palmerworms	3
Spiders	3
Scorpions	10
Worms	20
Total	**120**

List of References

Ants:	Proverbs 6:6, 30:25
Bees:	Deuteronomy 1:44, Isaiah 7:18, Judges 14:8, Psalms 118:12
Beetles:	Leviticus 11:22
Cankerworms:	Joel 1:4, 2:25; Nahum 3:15, 16
Caterpillars:	II Chronicles 6:28; Isaiah 33:4; Jeremiah 51:14, 27; Joel 1:4, 2:25; I Kings 8:37; Psalms 78:46, 105:34
Fleas:	I Samuel 24:14, 26:20
Flies:	Ecclesiastes 10:1, Exodus, 8:21, 22, 24, 29, 31; Isaiah 7:18; Psalms 78:45, 105:31
Gnats:	Matthew 23:24
Grasshoppers:	Amos 7:1; Ecclesiastes 12:5; Isaiah 40:22; Jeremiah 46:23; Job 39:20; Judges 6:5; 7:12, Leviticus 11:22; Nahum 3:17; Numbers 13:33; Psalms 105:35
Hornets:	Deuteronomy 7:20; Exodus 23:28; Joshua 24:12
Lice:	Exodus 8:16, 17, 18; Psalms 105:31
Locusts:	II Chronicles 6:28, 7:13; Deuteronomy 28:38, 42; Exodus 10:4, 12, 13, 14, 15, 19; Isaiah 33:4; Joel 1:4, 2:25; I Kings 8:37; Leviticus 11:22; Matthew

	3:4; Mark 1:6 Nahum 3:15, 17; Proverbs 30:27; Psalms 78:46, 105:34, 35, 109:23; Revelation 9:3, 7
Moths:	Hosea 5:12; Isaiah 50:9, 51:8; James 5:2; Job 4:19, 13:28, 27:18; Luke 12:33; Matthew 6:19, 20; Psalms 39:11
Palmerworms:	Amos 4:9; Joel 1:4, 2:25
Spiders:	Isaiah 59:5; Job 8:14; Proverbs 30:28
Scorpions:	II Chronicles 10:11, 14; Deuteronomy 8:15; Ezekiel 2:6; I Kings 12:11, 14; St. Luke 10:19, 11:12; Revelation 9:3, 10
Worms:	The Acts 12:23; Deuteronomy 28:39; Exodus 16:20, 16:24; Isaiah 14:11, 41:14, 51:8, 66:24; Job 7:5, 17:14, 19:26, 24:20; 25:6; Jonah 4:7; St. Mark 9:44, 46, 48; Micah 7:17; Psalms 22:6

Quotations

The Acts
12:23: And immediately the angel of the Lord smote him, because he gave not God the glory: and he was eaten by worms, and gave up the ghost.

Amos
4:9: I have smitten you with blasting and mildew when your gardens and your vineyards and your fig trees and your olive trees increased, the palmerworm devoured them: yet have you not returned unto me, saith the Lord.

7:1: Thus hath the Lord God shewed unto me; and, behold, he formed grasshoppers in the beginning of the shooting up of the latter growth; and, lo,it was the latter growth after the king's mowings.

II Chronicles
6:28: If there be dearth in the land, if there is pestilence, if there be blasting, or mildew, or locusts, or caterpillars; if their enemies besiege them in the cities of their land; whatsoever sore or whatsoever sickness there be:...

7:13: If I shut up heaven that there be no rain, or if I command the locust to devour the land, or if I send pestilence among my people:...

10:11: For whereas my father put a heavy yoke upon you, I will put more to your yoke: my father chastised you with whips, but I will chastise you with scorpions.

Deuteronomy

1:44: And the Amorites, which dwelt in that mountain, came out against you, as bees do, and chased you, and destroyed you in Seir, even unto Hormah.

7:20: Moreover the Lord Thy God will send the hornet among them until they that are left, and hide themselves from thee, be destroyed.

8:15: Who led thee through that great and terrible wilderness, wherein were fiery serpents, and scorpions, and drought, where there was no water; who brought thee forth water out of the rock of flint.

28:38: Thou shalt carry much seed out into the field, and shalt gather but little in for the locust shall consume it.

28:39: Thou shalt plant vineyards, and dress them, but shalt neither drink of the wine, nor gather the grapes; for the worms shall eat them.

28:42: All thy trees and fruit of thy land shall the locust consume.

Ecclesiastes

10:1: Dead flies cause the ointment of the apothecary to send forth a stinking savour; so doth a little folly him that is in reputation for wisdom and honour.

12:5: Also when they shall be afraid of that which is high, and fears shall be in the way, and the almond tree shall flourish, and the grasshopper shall be a burden, and desire shall fail; because man goeth to his long home, and the mourners go about the streets;

Exodus

8:16: And the Lord said unto Moses, Say unto Aaron, Stretch out thy rod, and smite the dust of the land, that it may become lice throughout all the land of Egypt.

8:17: And they did so; for Aaron stretched out his hand with his rod, and smote the dust of the earth, and it became lice in man, and in beast; all the dust of the land became lice throughout all the land of Egypt.

8:18: And the magicians did so with their enchantments to being forth lice, but they could not: so there were lice upon man, and upon beast.

8:21: Else if thou wilt not let my people go, behold I will send swarms of flies upon thee, and upon thy servants, and upon thy people, and into thy houses: and the houses of the Egyptians shall be full of swarms of flies, and also the ground whereon they are.

8:22: And I will sever in that day the land of Goshen, in which thy people dwell, that no swarms of flies will be there; to the end thou mayest know that I am the Lord in the midst of the earth.

8:24: And the Lord did so; and there came a grievous swarm of flies into the house of Pharaoh, and into his servants' houses, and into all the land of Egypt: the land was corrupted by reason of the swarm of flies.

8:29: And Moses said, Behold, I go out from thee, and I will intreat the Lord that the swarms of flies may depart from Pharaoh, from his servants, and from his people, tomorrow: but let not Pharaoh deal deceitfully any more in not letting the people go to sacrifice to the Lord.

8:31: And the Lord did according to the word of Moses; and he removed the swarms of flies from Pharaoh, from his servants, and from his people; there remained not one.

10:4: Else, if thou refuse to let my people go, behold, tomorrow will bring the locusts into thy coast.

10:12: And the Lord said unto Moses, Stretch out thine hand over the land of Egypt for the locusts, that they may come up upon the land of Egypt, and eat every herb of the land, even all that the hail hath left.

10:13: And Moses stretched forth his rod over the land of Egypt, and the Lord brought an east wind upon the land all that day, and all that night; and when it was morning, the east wind had brought the locusts.

10:14: And the locusts went up over all the land of Egypt: and rested in all the coasts of Egypt; very grievous were they; before them were no such locusts as they, Neither after them shall be such.

10:15: For they (the locusts) covered the face of the whole earth, so that the land was darkened; and they did eat every herb of the land, and all the fruit of the trees which the hail had left: and there remained not any green thing in the trees, or in the herbs of the field, through the land of Egypt.

10:19: And the Lord turned a mighty strong west wind, which took away the locusts, and cast them into the Red Sea; there remained not one locust in all the coasts of Egypt.

16:20: Notwithstanding they harkened not unto Moses; but some of them left of it (manna) until the morning, and it bred worms, and stank: and Moses was wroth with them.

16:24: And they laid it up till the morning, as Moses bade: and it did not stink, neither was there any worm therein.

23:28: And I will send hornets before thee, which shall drive out the Hivite, the Canaanite, the Hittite, from before thee.

Ezekiel
2:6: And thou, son of man, be not afraid of them, neither be afraid of their words, though briers and thorns be with thee, and thou dost dwell among scorpions: be not afraid of their words, nor be dismayed at their looks, though they be a rebellious house.

Hosea
5:12: Therefore will I be unto Ephraim as a moth, and to the house of Judah as rottenness.

Isaiah
7:18: And it shall come to pass in that day, that the Lord shall hiss for the fly that is in the uttermost part of the rivers of Egypt, and for the bee that is in the land of Assyria.

14:11: Thy pomp is brought down to the grave, and the noise of thy viols: the worm is spread under thee, and the worms cover thee.

33:4: And your spoil shall be gathered like the gathering of the caterpillar; as the running to and fro of locusts shall he run upon them.

40:22: It is He (God) that sitteth upon the circle of the earth, and the inhabitants thereof are as grasshoppers; that stretcheth out the Heavens as a curtain, and spreadeth them out as a tent to dwell in:

41:14:Fear not, thou worm Jacob, and ye men of Israel; I will help thee, saith the Lord, and thy redeemer, the Holy One of Israel.

50:9: Behold, the Lord God; will help me; who is he that shall condemn me? Lo, they all shall wax old as a garment; the moth shall eat them up.

51:8: For the moth shall eat them up like a garment, and the worm shall eat them up like wool; but my righteousness shall be for ever, and my salvation from generation to generation.

59:5: They hatch cockatrice eggs, and weave the spider's web; he that eateth of their eggs dieth, and that which is crushed breaketh out into a viper.

66:24: And they shall go forth, and look upon the carcasses of men that have transgressed against me; for their worm shall not die, neither shall their fire be quenched; and they shall be an abhorring unto all flesh.

James
5:2: Your riches are corrupted, and your garments are motheaten.

Jeremiah
46:23: They shall cut down her forest, saith the Lord, though it cannot be searched; because they are more than the grasshoppers, and are innumerable.

51:14: The Lord of hosts hath sworn by himself, saying, Surely I will fill thee with men, as with caterpillars; and they shall lift up a shout against thee.

51:27: Set ye up a standard in the land, blow the trumpet among the nations, prepare the nations against her, call together against her the kingdoms of Ararat, Minni, and Ashchenaz; appoint a captain against her cause the horses to come up as the rough caterpillars.

Job
4:19: How much less in them that dwell in houses of clay, whose foundation is in the dust, which are crushed before the moth.

7:5: My flesh is clothed with worms and clods of dust; my skin is broken, and become loathsome.

8:14: Whose (the hypocrites) hope shall be cut off, and whose trust shall be a spider's web.

13:28: And he, as a rotten thing, consumeth, as a garment that is motheaten.

17:14: I have said to corruption, Thou art my father; to the worm, Thou art my mother, and my sister.

19:26: And though after my skin worms destroy this body, yet in my flesh shall I see God:

21:26: They shall lie down alike in the dust, and the worms shall cover them.

24:20: The womb shall forget him; the worm shall feed sweetly on him; he shall be no more remembered; and wickedness shall be broken as a tree.

25:6: How much less man, that is a worm? and the son of man, which is a worm.

27:18: He buildeth his house as a moth, and as a booth that the keeper marketh.

39:20: Canst thou make him (the horse) afraid as a grasshopper? the glory of his nostrils is terrible.

Joel
1:4: That which the palmerworm hath left hath the locust eaten; and that which the locust hath left hath the cankerworm eaten; and that which the cankerworm hath left hath the caterpillar eaten.

2:25: And I will restore to you the years that the locust hath eaten, the cankerworm, and the caterpillar, and the palmerworm, my great army which I sent among you.

Jonah
4:7: But God prepared a worm when the morning rose the next day, and it smote the gourd that it withered.

Joshua
24:12: And I sent the hornet before you, which drove them out from before you, even the two kings of the Amorites; but not with thy sword nor with thy bow.

Judges
6:5: For they came up with their cattle and their tents, and they came as grasshoppers for multitude; for both they and their camels were without number: and they entered into the land to destroy it.

7:12: And the Midianites and the Amalekites and all the children of the east lay along in the valley like grasshoppers for multitude; and their camels were without number, as the sand by the seaside for multitude.

14:8: And after a time he (Sampson) returneth to take her (a wife), and he turned aside to see the carcass of the lion; and, behold, there was a swarm of bees and honey in the carcass of the lion.

I Kings
8:37: If there be in the land famine, if there be pestilence, blasting, mildew, locust, or if there be caterpillar; if their enemy besiege them in the land of their cities; whatsoever plague, whatsoever sickness there be;

12:11: And now whereas my father did lade you with a heavy yoke, I will add to your yoke; my father hath chastised you with whips, but I will chastise you with scorpions.

12:14: And spake to them after the counsel of the young men, saying, My father made your yoke heavy, and I will add to your yoke; my father also chastised you with whips, but I will chastise you with scorpions.

Leviticus
11:22: Even these of them ye may eat; the locust after his kind, and the bald locust after his kind, and the beetle after his kind, and the grasshopper after his kind.

Luke
10:19: Behold, I give unto you power to tread on serpents and scorpions, and over all the power of the enemy; and nothing shall by any means hurt you.

11:12: Or if he shall ask an egg, will he offer him a scorpion?

12:33: Sell that ye have, and give alms; provide yourselves with bags which wax not old, a treasure in the heavens that faileth not, where no thief approacheth, neither moth corrupteth.

Mark
1:6: And John was clothed with camel's hair; and with a girdle of skin about his loins; and he did eat locusts and wild honey.

9:44: Where their worm dieth not, and the fire is not quenched.

9:46: Where their worm dieth not, and the fire is not quenched.

9:48: Where their worm dieth not, and the fire is not quenched.

Matthew
3:4: And the same John had his raiment of camel's hair, and a leathery girdle about his loins; and his meat was locusts and wild honey.

6:19: Lay not up for yourselves treasures upon earth, where moth and rust doth corrupt, and where thieves break through and steal.

6:20: But lay up for yourselves treasures in heaven, where neither moth nor rust doth corrupt, and where thieves do not break through nor steal.

23:24: Ye blind guides, which strain at a gnat, and swallow a camel.

Micah
7:17: They shall lick the dust like a serpent, they shall move out of their holes like worms of the earth: they shall be afraid of the Lord our God, and shall fear because of thee.

Nahum
3:15: There shall the fire devour thee; the sword shall cut thee off, it shall eat thee up like the cankerworm: make thyself many as the cankerworms, make thyself many as the locusts.

3:16: Thou hast multiplied thy merchants above the stars of heaven: the cankerworm spoileth, and fleeth away.

3:17: Thy crowned are as the locusts, and thy captains as the great grasshoppers, which camp in the hedges in the cold day, but when the sun ariseth they flee away, and their place is not known where they are.

Numbers
13:33: And there we saw the giants, the sons of Anak, which come of the giants: and we were in our own sight as grasshoppers, and so we were in their sight.

Proverbs
6:6: Go to the ant thou sluggard; consider her ways, and be wise.

30:25: The ants are a people not strong, yet they prepare their meat in the summer.

30:27: The locusts have no king, yet go they forth all of them by bands;

30:28: The spider taketh hold with her hands, and is in kings' palaces.

Psalms
22:6: But I am a worm, and no man; a reproach of men, and despised of the people.

39:11: When thou with rebukes dost correct man for iniquity, thou makest his beauty to consume away like moth: surely every man is vanity.

78:45: He sent divers sorts of flies among them, which devoured them; and frogs, which destroyed them.

78:46: He gave also their increase unto the caterpillar, and their labour unto the locust.

105:31: He spake, and there came divers sorts of flies and lice in all their coasts.

105:34: He spake and the locusts came, and caterpillars, and that without number.

105:35: And (locusts and grasshoppers) did eat up all the herbs in their land, and devoured the fruit of their ground.

109:23: I am gone like the shadow when it declineth: I am tossed up and down as the locust.

118:12: They compassed me about like bees; they are quenched as the fire of thorns: for in the name of the Lord I will destroy them.

Revelation
9:3: And there came out of the smoke locusts upon the earth; and unto them was given power, as the scorpions of the earth have power.

9:7: And the shapes of the locusts were like unto horses prepared unto battle; and on their heads were as it were crowns like gold, and their faces were as the faces of men.

9:10: And they had tails like unto scorpions, and there were stings in their tails: and their power was to hurt men five months.

I Samuel
24:14: After whom is the King of Israel come out? After whom dost thou pursue? After a dead dog, after a flea.

26:20: Now therefore, let not my blood fall to the earth before the face of the Lord: For the King of Israel is come out to seek a flea, as when one doth hunt partridge in the mountain.

Used with permission from Bull. Entomol. Soc. Amer. 4(3): 75-78 (1958).

Are the Pyramids Deified Dung Pats?

Ilkka Hanski

One of the marvels of the insect world is the elaborate nesting behaviour of large dung beetles (for comprehensive reviews see Halffter and Matthews 1966 and Halffter and Edmonds 1982). Most of the interesting species live in tropical savannas or forests, away from the centres of population biology, which may partly explain why dung beetles, in spite of their highly evolved sexual cooperation and parental care, have been practically ignored by biologists other than dung beetle enthusiasts.

The large scarabs (family Scarabaeidae) can be divided functionally into two groups, the tunnellers and the rollers. The tunnellers dig a simple or a more complex tunnel system (Halffter and Edmonds 1982) directly below a dropping, into which they push dung for feeding or to start breeding. The rollers, either one sex alone or a pair of beetles together, first form a ball of dung then roll it some distance away before burying it into the soil. Fights about the balls are common (Matthews 1963): dung often in short supply, and even if it were not, a stolen ball means saved time in ball construction and perhaps, for a male, a won mate. Nest architecture varies widely, and nest may contain from one to tens of dung balls with an egg in each (Halffter and Edmonds 1982). It is well known that the ancient Egyptians paid particular attention to one dung beetle, *Scarabaeus sacer*, a widespread Mediterranean species, unmistakably depicted in numerous Egyptian writings and drawings. It has been more difficult for Egyptologists to explain the cause of this interest. The word *kheper* means 'to come to existence' and also signifies the scarab beetle, but no particular importance has been attached to this 'coincidence'. To understand the scarab's high esteem in ancient Egypt, one needs to know the biology of dung beetles as well as the Egyptian culture. A recent study by Yves Cambefort at the Natural History Museum in Paris combines knowledge of these two disciplines in a captivating study of beetles and people.

Cambefort first points out the great importance of cattle to the Egyptians, and how everything related to cattle was sacred and closely observed (as is still the case in some pastoral cultures). The ancient Egyptians made an association, evidenced by their writings, of the scarab beetle and the sun. The scarab's head resembles the rising sun. Beetles arrive at cow pats before sunrise, and continue to form and roll balls until early morning. After the ball has been rolled for some distance, it is buried in the soil, where a nest chamber is excavated. Beetles work in bisexual pairs, but mating occurs in the tunnel or in the breeding chamber and is rarely observed. Cambefort suggests that the Egyptians interpreted the breeding cycle as a union between the beetle and the earth.

The product of this union is the scarab larva inside the nesting ball. The nest of *S. sacer* represents a very simple type in dung beetles, usually consisting of a diagonal tunnel from the soil surface to the nesting chamber, in which a single nest ball is located. Following oviposition, the female leaves the nest and begins to construct another one. [In the related genus *Kheper*, also represented in Egypt, the female may stay in the nest during larval development, which necessarily leads to very low fecundity, down to the absolute minimum of a single offspring per breeding (rainy) season in *K. nigroaeneus* in southern Africa (Edwards 1984).]

At the end of its development, the larva turns into a pupa, and eventually a new scarab emerges. That acute observer of natural history, J-H. Fabre, was perhaps the first to liken the scarab pupa inside the nesting ball to a mummy (Fabre 1897). Cambefort turns Fabre's remark around and suggests that it was the scarab that gave the idea of mummification to the Egyptians. The scarab's nest becomes the burial chamber, in which the pupa-mummy awaits its solar resurrection. The dung ball was made out of a cow pat. One is not surprised that Cambefort draws the logical conclusion, even if tentatively, that pyramids are nothing but idealized cow pats.

With numerous pieces of evidence, Cambefort identifies the scarab with Osiris, the first mythical king of Egypt. Horus, the son of Osiris, is the new scarab that emerges from the pupa-mummy (Fig. 2). Another fascinating finding is a drawing of a scarab with the sign of *shen* behind the abdomen. Cambefort argues convincingly that the shen sign in this position represents the feces of the scarab. The scarab that eats the dung ball (the sun) ejects the sign of *shen*, 'the universe that the sun illuminates.' The same sign was used by the Egyptians as a frame for the names of their pharaohs and other royalty.

To appreciate the force of Cambefort's arguments one needs to read the original paper and, perhaps, be a competent Egyptologist, with a keen interest in dung beetles. For a population biologist, it is enthralling to observe the links between the highly evolved dung beetle breeding behaviour and one of the most remarkable of human cultures.

References

Cambefort, Y. 1987. Rev. de I'Hist. des Religions cciv-1, 3-46.

Edwards, P.B. 1984. XVII Int. Congr. Entomol., Abstracts, p. 338.

Fabre, J-H. 1897. Souven. Entomol. 5e, Serie, Paris.

Halffter, G. and Matthews, E.G. 1966. Folia Entomol. Mex. 12-14, 1-312.

Halffter, G. and Edmonds, W.D. 1982. The Nesting Behavior of Dung Beetles (Scarabaeinae), Instituto de Ecologia, Mexico, D.F.

Matthews, E.G. 1963. Psyche 70, 75-93.

Used with permission from Ecology and Evolution 3(2): 34-35 (1988)

Arthropods on the Screen

James W. Mertins

Recent rapid development of photographic and electronic technology plus growing environmental awareness have led to increased production of quality nature films. Very few now appear in theaters, but we can enjoy exceptional nature programming almost weekly on television through series such as *Nature, Nova, Survival,* and the *National Geographic Specials.* These efforts make a positive contribution to educating general viewers about nature, and some become available and useful for classroom purposes. But I wish to address the generally less constructive treatment afforded to arthropods on the screen in films made solely for commercial release. I present here a preliminary survey of the movies known to me whose production depicted or used arthropods in some way. Although primarily fictional creations of nonscientists, these works may serve to illustrate how one group of potential mass-audience "educators" (*i. e.* Hollywood film makers) views entomological subjects and how that view is transmitted by their medium.

For convenience I have divided the movies into three categories based on personal opinions of the importance of the entomological sequences to the essence of each film. In the first category, the entomological aspects are essential or very important to the plot or overall impact of each film; in category two, the arthropod/ entomology roles are, at best, subsidiary or incidental to the thrust of each movie. Some viewers might not agree with my somewhat subjective apportionment of certain films between the two categories. The third category is a list of films whose titles might infer some entomological involvement that is not really there. The films appear chronologically by year of release in the United States, and for each title (categories one and two only) a word or two is added about entomological content. I have personally seen many of the movies, but I have also examined all available contemporary and retrospective critical reviews to fill gaps in my knowledge.

The lists include only films that were considered to he of feature length at the time of release. Emphasis is on English language live-action or animated films released theatrically in the United States; however, some made-for-television movies are included because they were shown theatrically overseas. Except for British and Canadian productions, foreign movies are listed only if they were in general American release and reached a modicum of popularity (usually dubbed or subtitled in English). Not included are dozens of other foreign-language films, a few theatrically released documentaries, many older silent movies, an array of movie serials, and all sorts of television programs and short subjects.

Discussion

This is the first comprehensive review of the subject from an entomological perspective, although *Insectimes* (a quarterly trade publication of NOR-AM Chemical, Wilmington, DE) examined a few so-called "Big Bug Flix" beginning in 1981, and a chapter in Medved and Medved (1980) looked at movies about "killer bees" from the perspective of movie critics.

The Image of Entomology.

Several generalizations arise from the films listed in the first two tables (hereinafter called arthropod features). The first is that arthropod features rarely project positive images of arthropods, entomologists, or science in general. Footnotes mark exceptions, but arthropods generally present threatening, sinister images of danger or death; distasteful, shocking images for the squeamish; or obvious images of silliness for ridicule by any "rational" person. Entomologists (or usually the generic "scientists") are shown usually as detached from reality, as eccentric buffoons, often as psychotics, or, at least, as ineffectual dupes. One cannot find among the arthropod features a single instance in which scientific endeavor is shown in a completely positive light, and science or the scientific method are often misrepresented. According to Brustein (1958), "movies covertly embody certain underground assumptions about science which reflect popular opinions." But do the characters speak as representatives of the general public, or are the moviemakers speaking through them to influence viewers?

Movies marked by footnote *a* in the tables are those presenting a relatively positive or truthful image of arthropods (entomologically speaking), even though some might not be appealing to non-entomologists. That is, the insects may simply display normal behaviors of their respective ecological roles, fascinating to the entomologist but possibly disturbing to most other viewers. These images were probably included to project negative feelings. Not surprisingly, almost all the well-treated movie arthropods are at least somewhat anthropomorphized. The movies marked by footnote *b* in the tables are those that show entomologists in a predominantly positive or realistic way.

Quality of the Movies.

A second generalization about arthropod features is that Hollywood's use of arthropods is usually associated with the production of bad or mediocre movies. As a result, these films often play to limited audiences and many remain quite obscure. Major reasons for low quality are inadequate financing and simple incompetence. Of course, "bad" is a subjective based mostly on film critics' opinions, and in some cases films panned critically find success at the box office. Only rarely do arthropod features, especially those in category one, earn Oscar nominations. Particularly rare is the arthropod feature that wins a so-called "major" Oscar. All of the few arthropod features I have identified that have won major Oscars are in category 2; the most highly honored of these, *Annie Hall*, won three major Oscars plus one other in 1977. Typically, the minor roles of arthropods in these movies have little or no bearing on their honors. Not surprisingly, arthropod features are most often cited for their visual effects and for their musical scores, production aspects that often set or enhance an intended mood. Despite some awards, most arthropod features are abysmal failures, artistically, critically, and financially. It is difficult to describe how poorly written, badly acted, and cheaply made some of these films are. As just one example, in a nationwide Worst Film Poll (Medved and Medved 1980), two big-budget arthropod features placed among the top five, no. 2, *Exorcist II — The Heretic*, and no. 4, *The Swarm* (called by the Medveds "the most badly bumbled bee movie of all time"). In some cases (*e. g. Exorcist II*), the arthropod sequences may be the highest quality aspect of the movie.

Chronological Sequence.

An examination of arthropod features by decade reveals some trends. There seem to be some sizeable gaps in the recorded chronological appearances of arthropod features. These may be true reflections of periods when moviemakers neglected arthropod themes, or they may be artifacts of discontinuities in the filmographic reference materials available. During the silent film and early talkie eras, there were very few feature movies with a major entomological theme. The early arthropod features in category two used entomological themes and images in ways divided about equally between positive and negative. Finally, many silent movies ascribed assumed arthropod characteristics to a leading character, and, hence, the name of the arthropod was metaphorically given to the character and often the movie. Such arthropodization of human characters is the unifying feature of nearly all the films listed in the third table. Movies with significant entomological content seem curiously absent during the 1930s.

The 1940s began benignly with three entertaining films featuring lovable anthropomorphic insects, but sinister specters (especially bees and spiders) filled most of the few other notable films of the decade.

The 1950s also began calmly, with enjoyable but brief roles for caterpillars in *Alice in Wonderland* and *Hans Christian Andersen*. But there was almost nothing entomologically positive in arthropod features for the rest of the decade, and although many films were made, the 1950s were the low point for entomology in the movies. *Them!* (1954) was the pivotal film in the history of arthropod features and, in fact, for the entire fantasy movie genre. It set a plot pattern that was followed frequently: scientific tampering with the unknown could often loose awful consequences on humankind. There were some pretty entertaining arthropod features in the 1950s, but most emphasized negative aspects of arthropods and were incredibly bad.

A few positive images occur in arthropod features of the 1960s, but mostly the prejudices of the 1950s continued. Nearly all of the bright spots occur in films from Japan. The other movies of the 1960s that reflect well on arthropods show them as pets of some sort. Many of the remaining features of the 1960s variously reworked the old mad scientist or science generated monster themes or portrayed arthropods as agents of injury and death.

The greatest array of arthropod features appeared in 1970s, but the positive aspects were mostly restricted to youth oriented cartoon characters. A few entomologists were positively portrayed, but most of the others were psychotic or bizarre. The physical size of 1970s arthropods (villainous and otherwise) tended toward more realistic proportions, although a few giant crabs, spiders, cockroaches, and Hymenoptera turned up. Moviemakers must have kept up on news reports, for when the presence of the African honey bee was first publicized widely (Anonymous 1972), there quickly began a series of "killer bee" movies The production credits to these and some other 1970s arthropod features provide a possible clue to the shifting image of arthropods on the screen. Directors began to hire and give credit to entomological consultants (sometimes real entomologists and often called "insect wranglers") to provide and manipulate arthropods according to the scripts. This increased alleged sophistication and (one hopes) rational advice from the consultants (advice, unfortunately, not always followed) may explain why the frightening, but impossibly overgrown, giant arthropods all but disappeared in 1970s films.

Thus far, arthropod features in the 1980s seem to be extending the late 1970s tradition. Fewer films are using arthropods in major themes, but more are using them in subsidiary ways. Arthropods are being used more intelligently and in more realistic ways. The emphasis seems to be more on the presence and activities of arthropods as contributory realistic details.

Taxonomic Frequencies.

For various reasons, it is not always easy to identify screen arthropods with assurance. However, identities can be fixed well enough to show that arachnids clearly appear most frequently. Except for a few scorpions, the majority of arachnids used are spiders, and tarantulas are the favorites. The spider image invariably represents evil and morbidity, and the film makers, relying on the viewers' learned fears or innate fear of the unknown, seem to believe in the effectiveness of spiders for conveying or creating a dark and sinister mood. If audience fear of pain, injury, or death from encounters with arachnids is the key to the popularity of spiders and scorpions, then the Hymenoptera must be second most popular for similar reasons, headed by 17 appearances by bees. Next, the Lepidoptera are mostly treated neutrally or with favor, but even they sometimes play killers (e. g., *Mothra, The Vampire Beast Craves Blood, The Butterfly Murders*). A few other taxa make multiple but less common appearances, including orthopteroids, crustaceans, and Diptera, but the relative rarity of beetles in films is notable in light of the great number of species.

Other Observations.

Hollywood often relies on arthropods to stimulate audience emotions and to create a desired mood. A number of moviemaking devices involving arthropods are used frequently to amplify a desired response. First there is the minor but interesting idea of adding cricket chirping to the soundtracks of countless films during nighttime scenes. Similarly, dipteran buzzes or slaps at phantom "bugs" may represent pestilence and irritation. A fairly subtle means of heightening the tension between arthropod and human characters is the inevitable meeting of the two in a closed space from which there is no easy escape, places like cellars, locked rooms, islands, and, especially, tunnels, caves, and caverns. Another dramatic device is a much more blatant play for a quick, graphic shock reaction. It is the repetitive use of entomophagy by humans to jar the audience; this effort to disgust has been used frequently, especially in the last decade. Tactile contact with walking arthropods on human skin is similarly exploited in many movies, as in the uncounted westerns, for example, wherein people are buried or staked with bare skin exposed to torture by ants. Actors walking through cobwebs is a similarly exploited situation. If one watches carefully insects sometimes appear on screen unintentionally, especially in the backgrounds of movies filmed on location in jungles or other tropical locales. Finally, Steven Kutcher (personal communication), a practicing entomological consultant to Hollywood film makers, believes that 25-35% of all feature films have some entomological reference in them, even though some may be as minor as a picture on a wall or a single word of incidental dialogue.

Conclusion

Brustein's (1958) contention that movies and film makers reflect society's opinions about science seems to apply to a degree specifically to entomology. Moviemakers sense public opinions about their films through test audiences and especially by box office receipts. Their primary aim is to make a profit and, in the process, to entertain. They know how to tweak an audience, what succeeds, and what sells. Audience distaste for arthropods is a proven filmmaking ploy, and, thus, most arthropod appearances on screen are for scares in so-called "exploitation" (*e. g.,* horror, fantasy, and science fiction) films. If a film entertains and rings true with audiences, it succeeds, and it may inspire more like it. But Brustein also inferred that film makers judge scientific endeavors and furtively include their own feelings in their movies, thereby not only reflecting public opinion but also serving to form and direct it. Commercial films have a very large potential audience that mostly consists of impressionable young people. Unfortunately, the movie images of arthropods and entomology shown to them have been mostly inaccurate and unflattering.

One previous study dealt with another potential medium of mass education about entomology (Moore *et al.* 1982). Recent magazine articles and pictures aimed at the general public show a strong bias against insects in articles of an entomological nature. Except in nature magazines, insects are shown mostly in an unfavorable light as "bad organisms." And it seems that the movie image problem for arthropods and entomology has been exactly comparable. However, one hopes the recent trend continues, wherein arthropod features are of higher quality, more realistic, and more intelligent than in the past. The change may result from the increasing help of entomological consultants, or it may be from a general maturation of attitudes, ideas, and abilities in audiences and the film industry. I hope the trend is real, but I must point out that 1984 brought us *Runaway,* a movie with a new twist, the first robotic arthropods, and Hollywood made them evil! Let us hope that this sort of portrayal is a short-lived fad.

Acknowledgment

I express my gratitude to May Berenbaum for early interest and suggestions, to Steven Kutcher for much useful information and insight, and thanks to Marilee Menins for buying so many movie tickets for me.

References Cited

Anonymous. 1972. Final report, Committee on the African Honey Bee. National Academy of Sciences, Washington, D.C.

Brustein, R. 1958. Reflections on horror movies. Partisan Rev. 25: 288-296.

Medved, M. and H. Medved. 1980 The golden turkey awards nominees and winners — the worst achievements in Hollywood history. Putnum, New York.

Moore, W. S., D. R. Bowers and T. A. Granovsky. 1982. What are magazine articles telling us about insects? Journalism Quarterly 59: 464-467.

Used with permission from Bull. Entomol. Soc. Amer. 32(2): 85-90 (1986).

Table 1. Movies with entomological elements of significant importance, including films with positive, or at least neutral, realistic images of arthropods ([a]) and those with generally positive images of entomologists ([e]).

1908   *Those Flies* (British: silent; a.k.a. *The Flies*) - flies.

1911   *The Fly's Revenge* (British; silent). fly.

1914   *Bumbles Goes Butterflying* (British; silent) - butterfly.

1940   *Pinocchio*[a] - anthropomorphic cricket; animated.

1941   *Hoppy Goes to Town*[a] (a.k.a. *Mr. Bug Goes to Town* - anthropomoprhic all-arthropod cast; animated.

1944   *Once upon a Time*[a] - dancing caterpillar. *Bees in Paradise* (British) - beelike island society. *The Spider Woman* (British; a.k.a. *Sherlock Holmes and the Spider Woman*) -spider.

1947   *The Black Widow* (serial; later [1946] recut as a feature movie, *Sombra, the Spider Woman*) - spider.

1954   *The Naked Jungle* - army ants. *Them!* - giant ants.

1955   *Tarantula* - giant tarantula. *Panther Girl of the Kongo* (serial; later [1966] recut as a TV movie, *The Claw Monsters*) - enormous crayfish.

1957   *The Monster that Challenged the World* - mollusca/annelid/caterpillar hybrid. *The Black Scorpion* - giant scorpions and trapdoor spider. *The Deadly Mantis* - giant preying mantis. *The Beginning of the End* - giant grasshoppers. *Attack of the Crab Monsters* - giant crab.

1958   *The Monster from Green Hell* - giant wasps. *The Cosmic Monsters* (a.k.a. *Cosmic Monsters*) - giant spiders and insects. *The Spider* (a.k.a. *Earth vs the Spider*) - giant tarantula. *The Fly* - fly and spider.

1959   *The Wasp Woman* (a.k.a. *The Bee Girl* or *Insect Woman* - wasps. *Return of the Fly* - fly. *The Tingler* - centipedelike beast (with "Percepto" effects).

1962   *Mothra* (Japanese) - giant silkworm/cocoon/moth. *It's Hot in Paradise* (German; a.k.a. *Hot in Paradise, The Spider's Web,* or *Girls of Spider Island*; rereleased in 1965 as *Horrors of Spider Island*) - huge spider.

1964   *Godzilla vs. the Thing*[a] (Japanese) - the "thing" is reformed Mothra with two giant larval offspring.

1965   *The Collector* (British) - demented amateur lepidopterist.

1967   *The Deadly Bees* (British; a.k.a.*KillerBees*) - beekeeper and trained bees. *Silence Has No Wings*[a] (Japanese) - rare butterfly and schoolboy collector.

1969   *Kenner*[a] - cricket. *Godzilla vs. the Sea Monster* (Japanese; a.k.a. *Ebirah, Horror of the Deep*) - giant crustacean and Mothra. *The Vampire Beast Craves Blood* (British; a.k.a. *The Blood Beast Terror, The Deathshead Vampire,* or *Blood Beast from Hell*) - entomology professor, students, and giant death's-head sphinx.

1970   *Flesh Feast* - maggots.

1971   *The Hellstrom Chronicle* - semidocumentary with close-up insect photography; *The Legend of Spider Forest* (British, a.k.a. *Venom*) - spiders.

1972   *Kiss of the Tarantula* (a.k.a. *Shudders*) - tarantualas.

Table 1. Continued,

1973   *Charlotte's Web*[a] - anthropomorphic spider, animated. *Invasion of the Bee Girls* (a.k.a. *Graveyard Tramps*) - deranged entomologist, mutant superwomen/queen bees, apiculture minicourse; softcore porn.

1974   *Locusts* (TV) - plague of locusts. *Phase IV* - intelligent ants. *The Killer Bees* (TV) - bees.

1975   *Bug* - large, potent, intelligent cockroaches.

1976   *The Giant Spider Invasion* - tarantula hordes. *Godzilla vs. Megaton* (Japanese) - giant, anthropomorphic, quadrupod scarab beetle. *The Butterfly Ball*[a] (British) - fairy tale-based rock concert with animated insects. *The Savage Bees* (TV) - Africanized honey bees.

1977   *The Autobiography of a Flea* - voyeuristic talking flea; hardcore porn. *Tarantulas: the Deadly Cargo* (TV) - tarantulas. *Empire of the Ants* - minicourse in formicology plus giant ants. *Kingdom of the Spiders*[o] - tarantula hordes. *Curse of the Black Widow* (TV; a.k.a. *Love Trap*) - spiders. *It Happened at Lakewood Manor* (TV; a.k.a. *Panic at Lakewood Manor*) - killer ants. *Spider-Man*[a] (TV; a.k.a. *The Amazing Spider-Man*) - spider abilities. *The Swarm*[o] - Africanized honey bees. *Terror Out of the Sky* (TV) - Africanized honey bees. *The Bees* - Africanized honey bees. *Dr. Scorpion* (TV) - spiders and scorpions. *Spider-Man Strikes Back*[a] (TV) - spider powers.

1979   *Spider-Man the Dragon's Challenge*[a] (TV) - spider powers. *Alien* - environmentally resistant operculate egg, crustacean-like creature, ecdysis, exuviae, hypermetamorphosis. *The Butterfly Murders* Hong Kong) - butterfly swarms.

1980   *Island Claws* (a.k.a. *The Nigh of the Claw* - sand crab hordes plus palm crab.

1982   *Halloween III: Season of the Witch* - cockroaches, ants, spiders, etc. *Creep Show* - cockroach hordes.

1983   *A Swarm in May*[a] (British) - honey bee swarm.

1984   *Where the Green Ants Dream*[ao] (German) - entomologist and visionary ants. *Runaway* - hexapodous robots.

Table 2. Movies with relatively minor entomological elements, including films with positive, or at least neutral, realistic images of arthropods ([a]) and those with generally positive images of entomologists ([o]).

1916   *The Tarantula* (silent) - tarantula.

1918   *The Fly God*[a] (silent) - house fly.

1920   *The Beetle* (British; silent) - scarab beetle.

1923   *The Cricket on the Hearth*[a] (silent) - cricket.

1924   *Never Say Die* (silent) - bee.

1925   *The Keeper of the Bees*[ao] (silent) - beekeeper, apiary; talkie version remakes in 1935 and 1947. *Blue Blood*[o] (silent) - athletic entomologist.

1927   *Swim, Girl, Swim* (silent) - collegiate entomology and professor.

Table 2. Continued.

1928   *Eva and the Grasshopper* (German; silent) - anthropomorphic mechanical insect characters.

1930   *All Quiet on the Western Front*[a] - butterfly.

1937   *The Good Earth* - plague of locusts.

1940   *Dr. Cyclops* - looney biologist with butterfly net. *The Thief of Bagdad* - giant spider.

1943   *Tarzan's Desert Mystery* - giant spider.

1951   *Alice in Wonderland*[a] - caterpillar character; animated. Caterpillar also appears in many of the screen versions of this Lewis Carroll tale (1910, 1915, 1931, 1933, 1951, 1972, and 1976). *The African Queen* - swarms of hematophilous Diptera (possibly mosquitoes).

1952   *Hans Christian Andersen*[a] - "inchworm." *Mesa of Lost Women* (a.k.a. *Lost Woman* or *Lost Womenof Zarpa*) - hugh tarantulas.

1953   *Cat Women of the Moon* (3-D; a.k.a. *Rocket to the Moon*) - hugh horned spiders. *Port Sinister* (a.k.a *Beast of Paradise Island*) - oversized crabs. *It Came from Outer Space* (3-D) - metaphoric spiders.

1954   *Killers from Space* - large insects.

1956   *Congo Crossing* - horde of tsetse flies. *World without End* - giant spiders.

1957   *She Devil* - "miracle" fruit fly serum. *The Incredible Shrinking Man* - "huge" spider.

1958   *Rodan* (Japanese; a.k.a. *Rodan, the Flying Monster*) - large prehistoricdragonflies (*Meganeura*). *Missle to the Moon* (remake of *Cat Women of the Moon*, 1953) - hairy moonspiders. *Monster on the Campus* - big dragonfly. *Queen of Outer Space* - giant spider.

1959   *Have Rocket, Will Travel* - the Three Stooger and a giant tarantula. *Tarzan's Greatest Adventure* (British) - tarantula. *Teenagers from Outer Space* (a.k.a. *The Gargon Terror*) - giant crustacean. *The Angry Red Planet* - chimeric spiderlike monster.

1960   *The Lost World* (a.k.a. *The Origin of Man*) - large spiders.

1961   *Valley of theDragons* (a.k.a. *Prehistoric Valley*) - giant spider. *Mysterious Island* - one of five film versions of this tale, this one has giant land crabs and deadly bees.

1962   *Mr. Arkadin* (British; a.k.a. *Confidential Report*) - flea circus.

1963   *Dr. No* - 007 meets a tarantula. *Summer Magic*[a] - song illustrated by Disney nature footage of various insects. *Captain Sinbad* - symbolic spider-and-fly dance routine.

1964   *The Mighty Jungle* - scorpion and tarantula fight. *Woman in the Dunes*[o] (Japanses) - entomologist/collector. *First Men in the Moon* (British) - giant caterpillar and insectlike humanoids.

1965   *Ghidrab, the Three-headed Monster*[a] (Japanese; a.k.a. *The Biggest Fight on Earth*) - Mothra's caterpillar offspring. *Sands of the Kalahari* (British) - migratory locust swarm. *Pinocchio in Outer Space* - giant crabs; animated. *Spider Baby* (a.k.a. *TheMaddest Story Ever Told* or *The Liver Eaters*) - spiders. *Space Monster* (a.k.a. *First Woman into Space* or *Voyage Beyond the Sun*) - giant crabs. *The Lost World of Sinbad* (Japanese; a.k.a. *Samurai Sailor* or *Sinbad the Sailor*) - flying insect, formerly a witch.

Table ? Continued.

1966   *Crazy Quilt* - exterminator. *The Appaloosa* - scorpions. *Let's Kill Uncle* -
       tarantula. *Women of the Prehistoric Planet* - giant spiders. *Dead Heat on
       a Merry-Go-Round* - termite exterminator. *One Million Years B. C.*
       (British) - giant spider.
1967   *Doctor Dolittle*[a] - "Giant Lunar (sic) Moth."
1968   *The Lost Continent* (British) - hugh crabs. *The Name of the Game is Kill*
       (a.k.a. *Lovers in Limbo*) - spiders. *Son of Godzilla* (Japanese) - giant
       preying mantids and a giant spider. *Wild in the Streets*[a] - crayfish.
1969   *Destroy All Monsters*[a] (Japanese; a.k.a. *Operation Monsterland*) - Mothra,
       giant spider, giant mantis, etc. *Five Bloody Graves* (a.k.a. *The Gun Riders,
       The Lonely Man,* or *Five Bloody Days to Tombstone*) - man staked to
       anthill. *Some Kind of Nut* - bee.
1970   *Birds Do it, Bees Do It*[a] - mantids, maggots, etc.
1971   *The Abdominable Dr. Phibes* (a.k.a. *Dr. Phibes*) - bees and locusts.
       TwoLaneBlacktop - chorusing cicadas. *When Dinosaurs Rules the Earth*
       (British) - giant crabs. *Yog - Monster from Space* (Japanese; a.k.a. *Space
       Amoeba*) - giant crustacean.
1972   *Frogs* - amateur lepidopterist, tarantulas, butteflies, and crabs.
1973   *Papillon* - cockroaches, centipedes, entomorphagy.
1975   *The Devil's Rain* - tarantula.
1976   *Food of the Gods* - oversized wasps and caterpillars. *The Missouri Breaks*
       - moth; mention of scale insects and apple-tree borers.
1977   *Damnation Alley* - large cockroaches, giant scorpions. *Annie Hall* -
       lobsters nad (unseen) spider "big as a Buick."
1978   *Planet of Dinosaurs* - spiders. *Exorcist II: The Heretic* - demonic migra-
       tory locusts.
1979   *Dracula* - cockroaches, entomophagy. *Human Experiments* (a.k.a. *Beyond
       the Gate*) - cockroaches, spiders, etc. *The Amityville Horror* - swarms of
       demonic house flies.
1980   *Urban Cowboy* - giant skipper larva.
1981   *Raiders of the Lost Ark* - cave full of tarantulas. *Clash of the Titans* - huge
       scorpions. *The Dogs of War* - cockroach.
1982   *The House Where Evil Dwells* - ghostly plague of insects and possessed
       big crabs. *Quest for Fire* - moth, entomophagy, lousing behavior. *The
       Sender* - cockroaches. *Victor Victoria* - cockroaches. *It Came from
       Hollywood* - anthology of film clips including seven arthropod features.
1983   *Strange Brew* - moths. *Gorky Park*[a] - dermestid larvae. *Legent of the
       Champions* - locusts.
1984   *Iceman* - entomophagy. *The Karate Kid*[a] - house fly. *Greystoke: The
       Legend of Tarzan, Lord of the Apes* - European explorer/entomologist,
       lousing behavior, termites, entomophagy. *Indiana Jones and the Temple
       of Doom* - scarab beetles, entomophagy, cockroaches, phasmids,
       centipedes, cerambycids, house fly. *Irreconcilable Differences* - cage full
       of flies, insect nets. *The Terminator* - blow flies. *The Ewok Adventure*
       (TV) - huge spiders. *The Cotton Club* - house fly.

Table 3. Movies with no known true entomological connection.

1915, 1932, 1956 *Madame Butterfly.*
1915   *Fanchon the Cricket: The Scorpion's Sting; A Butterfly on the Wheel; The Moth and the Flame.*
1916, 1952 *The Spider and the Fly.*
1916, 1931, 1939, 1945 *The Spider.*
1917   *The Crab; Lady Barancle; The Fly Crop; The Firefly of Tough Luck; The Desire of the Moth; The Cricket.*
1918   *The Wasp; The Firefly of France.*
1919, 1931, 1955 *Daddy Long Legs.*
1920   *The Black Spider; The Butterfly Man.*
1921   *Heedless Moths; The Butterfly Girl; The Scarab Ring.*
1922   *Butterfly Ranch.*
1923   *The Spider and the Rose.*
1924   *Sting of the Scorpion; The White Moth.*
1925, 1981 *Butterfly*
1925   *Broadway Butterfly; The Golden Cocoon.*
1926   *Butterflies in the Rain; The Gilded Butterfly.*
1927   *The Ladybird; Spider Webs.*
1928   *Black Butterflies.*
1931   *The Sky Spider.*
1934   *The Moth.*
1937   *The Firefly.*
1941, 1954   *Black Widow.*
1946   *Spider Woman Strikes Back.*
1951   *The Black Widow; The Sea hornet.*
1954   *Dragonfly Squadron.*
1955   *The Cobweb; Queen Bee.*
1957   *Joe Butterfly.*
1962   *Satan Bug.*
1963   *Lord of the Flies; Ladybug, Ladybug.*
1965   *Scream of the Butterfly; Curse of the Fly.*
1968   *A Flea in Her Ear.*
1969   *The Love Bug; Along Came a Spider (TV); The Gypsy Moths; Float Like a Butterfly; Sting Like a Bee.*
1970   *Hornet's Nest; The Grasshopper; Mosquito Squadron.*
1972   *Super Fly; Butterflies Are Free.*
1975   *Day of the Locust.*
1976   *Dragonfly.*
1980   *The Exterminator.*
1983   *Ziggy Stardust and the Spiders from Mars.*
1984   *Exterminator 2; Exterminators of the Year 3000.*

Stranger than Fiction

James H. Trosper

Most peoples' familiarity with the insect world is limited to the destructiveness they cause as competitors for our food and fiber and as the source of painful bites and stings. It is therefore hardly surprising that people so often have negative opinions about insects. The abnormal fear of insects (entomophobia) is one of the most common phobias, no doubt accounting for the widespread philosophy that the only good bug is a dead bug. This is tragic because insects do far more good than harm (less than 3% of known species are regarded as injurious), and their varied ways of life can be an endless source of wonder and delight. Few appreciate the amazing diversity of insect life. The insects in the average backyard are numerous and varied enough to occupy one's attention for a lifetime. As the 19th century scientist Louis Agassiz said "I spent the summer traveling. I got halfway across my backyard." However the diversity in your yard is a pale reflection of the abundant and bizarre insect life of the tropics. About one million different insects have been described and catalogued by entomologists so far. However, recent work by scientists in the canopy of tropical rain forests has produced startling estimates that possibly 30 million insect species may exist — far higher than previous estimates. With such incredible diversity, it is not surprising that insects have established themselves in virtually every habitat, from snowfields high in the Himalayas to mines over a mile deep, from the turbulent surface of the ocean to pools of petroleum oil. The endless adaptability of insects amazes even professional entomologists who spend their lives studying them.

"Compared with ourselves, insects are peculiarly constructed animals. They might be said to be inside out because their skeleton is on the outside, or upside down because their nerve cord extends along the lower side of the body and the heart lies above the alimentary canal. They have no lungs, but breathe through a number of tiny holes in the body wall — all behind the head — and the air entering these holes is distributed through the body and directly to the tissues via a multitude of tiny branching tubes. The heart and blood are unimportant in the transport of oxygen to the tissues. Insects smell with their antennae, some taste with their feet, and some hear with special organs in the abdomen, front legs, or antennae" (Borror, et al. 1981). The mechanics of support and growth in an animal whose skeleton is on the outside of the body limit insects to a relatively small size. The giants of the insect world include walking sticks a foot long in Malaysia and the Atlas moth of Australia with a wingspread of 14 inches. These are the exceptions; most insects are less than a quarter of an inch long. However there are fossil records of dragonflies with wingspans in excess of 26 inches. We can be glad we weren't around at that time if the mosquitoes these dragonflies preyed upon were correspondingly large.

Even today's blood sucking insects can take an enormous blood meal relative to their size. The kissing bugs of Central and South America ingest a blood meal over 12 times their body weight. That is the equivalent of a man drinking over 200 gallons of beer at one sitting. Not even many alcoholics can achieve that feat.

Insects make the job of a special effects person for horror films easy. Close-ups of their complex mouthparts, pitted bodies, spines, hairs, and unusual appendages are enough to cause revulsion in the average person. Their bizarre shapes are certainly grist for a fertile imagination. Some snout beetles have the front of the head drawn out into a slender structure longer than the rest of the body but with tiny jaws at the end. Stalk-eyed flies have their eyes situated at the ends of long slender stalks which may be as long as the wings. Many stag beetles have jaws half as long as the body and branched like the antlers of a stag. Such morphological peculiarities are endless.

Even more amazing than their unusual shapes and structures are the astonishing things insects can do. Some fly larvae can tolerate the hot water in thermal springs where they feed on heat-resistant algae. Other species can survive long periods of subfreezing temperatures. Insects feed upon almost every organic substance found in nature, a fact reflected by their diverse digestive systems. These catholic tastes in food can cause some unusual problems for man. A species of beetle in the family Bostrichidae that occurs in the western United States normally bores in the wood of trees. Occasionally it bores into the lead sheathing of telephone cables allowing moisture to enter and causing a short-circuiting of the wires with consequent interruption of service. Not surprisingly, this insect has acquired the common name "lead cable borer." Although most people are aware of the fact that mosquitoes obtain blood for food, few are familiar with some unusual feeding habits found in this group. Mosquitoes of the Old World genus *Malaya* obtain their food by stroking certain species of ants with their antennae to entice them to regurgitate their stomach contents. And blood feeding is by no means confined to the familiar groups that plague us. Certain moths in Southeast Asia feed on blood in addition to lachrymal secretions around the eyes.

The reproductive powers of insects are awesome. "The fruit fly that has been studied so much by geneticists develops rapidly and under ideal conditions may produce 25 generations in a year. Each female will lay up to 100 eggs, of which about half will develop into males and half into females. Now suppose we start with a pair of these flies and allow them to increase under ideal conditions, with no checks on increase, for a year — the original and each succeeding female laying 100 eggs before she dies and each egg hatching, growing to maturity, and reproducing again, at a 50:50 sex ratio. From two flies in the first generation, would give rise to 100 in the second, 5000 in the third, and so on. By the twenty-fifth generation, there would be about 1.192×10^{41} flies. If this many flies were packed tightly together, 1000 to a cubic inch, they would form a ball of flies 96,372,988 miles in diameter-or a ball extending approximately from the earth to the sun!" (Borror, *et al.* 1981) Small wonder insect pests can be so difficult to control.

"Throughout the animal kingdom an egg usually develops into a single individual. In man and some other animals an egg occasionally develops into two individuals (for example, identical twins in man) or, on rare occasions, three or four. Some insects carry this phenomenon of polyembryony (more than one young from a single egg) much further. Among encyrtid wasps, over 1000 young may develop from a single egg. A few insects employ another unusual method of reproduction: paedogenesis or reproduction by larvae." (Borrer, *et al.* 1981) Such reproductive versatility has resulted in some unusual modifications of the reproductive system. Females of many species possess a saclike structure called the spermatheca in which sperm from the male are stored. Frequently a single mating is sufficient to fertilize all the eggs to be laid during the lifetime of the insect. Think of the children you would have to support if your wife had a spermatheca! Nearly all the higher insects copulate. Variations in the number of positions utilized brings to mind an insectan Kama Sutra.

"The mating act of Odonata is one of the most bizarre performances to be seen anywhere. Doubtless if a trip to the Kalahari desert were required to observe it, more publicity would result. But the fact is that a short time spent in quiet observation by pond or stream in the summer is all that is needed. Before discussing mating, it is necessary to describe the abdomen briefly. There is nothing special about the female, the long, needle-like abdomen merely terminating in the sexual orifice and the egg-laying apparatus. In the male, however, the abdomen terminates in a double set of powerful clasping organs-not unusual in itself until one realizes that these organs are not designed for grasping the genital organs of the female but for grasping her neck. In fact, many female dragonflies have grooves and pits on the back of the head into which these claspers fit. Here then is a dilemma. The claspers (and genital opening) of the male are located at this tail end, but this is applied to the neck of the female rather than to her genital opening. The dilemma is solved by the possession, in these and in no other insects, of a special genital pouch beneath the forward segments of the male abdomen — actually well in front of the middle of the body. Here one finds a jointed penis and a variety of hoop-like appendages associated with a sac that has no connection with the testes at all. Thus the male must fill these secondary genital organs, which he does by making a loop of his abdomen, applying the tip briefly to the underside of the genital pouch, and permitting sperm to enter the sac at the base of the penis. In damselflies, the sac is often filled after the male has grasped the female, but most dragonfly males are 'loaded' before they pair with a female." (Evans 1978)

"The act of copulation is initiated by the male. Following slightly above the female, he seizes her thorax with his legs. At this point, if the female is receptive, she flies slowly and allows herself to be carried. The male next makes a loop of his abdomen and grasps the female in the neck region with his claspers. He then lets go with his legs, straightens his abdomen, and flies ahead, carrying the female in tandem. The female normally makes a loop downward and forward with her abdomen and applies her genital opening to the genital sac of the male. This is called the "wheel position," since in fact the two form a loop, the female mostly upside-down beneath the male. The pair may remain together for anywhere from three seconds up to an hour or more, but all species that require more than a few seconds copulate while perched." (Evans 1978)

Psychologists might be interested to know that sadomasochism extends to the insect world. "Mating in the bedbug has been termed 'traumatic insemination,' meaning simply that the male punctures a hole in the female abdomen and inserts his semen there. His intromittent organ is a stout hook, suggesting half of an ice tongs. The hook is inserted through the membrane between the fifth and sixth abdominal segments of the female. Here a small notch guides the male penis into a swollen mass of tissue called the 'organ of Berlese'. There is, however, no opening to the outside, and the male actually punctures the body wall of the female. After mating, the wound heals over and a scar forms. The female can keep no personal secrets from the entomologist, who only has to count the scars to find out how often she had mated." (Evans 1978)

"The male bed bug injects an unusually large amount of semen into the female. The organ of Berlese apparently acts as something of a pad to protect the internal organs from laceration. It also helps to prevent bleeding, assists the healing of the wound, and absorbs much of the seminal fluid. The sperm make their way through the blood to the true reproductive organs of the female, finally gathering in little sacs called sperm reservoirs. Here they remain until the female takes a meal, whereupon they migrate up the egg tubes to the ovaries, ready to fertilize the eggs that are rapidly fabricated from the nutrients in the blood of the host." (Evans 1978)

Some insects have sanitized the sex act in a manner that would meet with Victorian approval. Males of some species enclose their sperm in a capsule called the spermatophore that is deposited on the ground. The female finds the spermatophore and inserts it into her reproductive opening to fertilize the eggs.

Insects face many challenges in meeting the conditions necessary for their existence. Those that live in water get their oxygen in unusual ways. Some mosquito larvae insert the posterior end of the body containing the openings to the respiratory system into the air spaces of submerged aquatic plants. Many aquatic bugs and beetles form their own diving bell by carrying a thin film of air on the surface of the body when they submerge. Dragonfly nymphs breathe by drawing water into the rectum through the anus and then expelling it. Expulsion of water from the anus also serves as a means of locomotion, allowing nymphs to escape from enemies by a form of jet propulsion.

Life on the water surface is no less fantastic. The surface tension of water can bear considerable weight. The feet of water striders are equipped with short waxy hairs that allow these insects to stand and scull on the surface film. Try to catch one and you find how quickly they can move on this fragile surface. These insects keep alert for vibrations from potential prey which could mean dinner has fallen on the water surface. Other surface vibrations serve to communicate between water striders, to attract males to females, or to establish territories between these frequently cannibalistic insects. Some water striders ride the surface of the open ocean and lay their eggs on rafts no more substantial than floating feathers dropped by birds. A few insects like the small elmid beetles are able to exploit the surface tension from below and walk upside down on the underside of the film, much like a fly walking on the ceiling. The whirligig beetles that dazzle observers with their wild gyrations swim at the surface with the upper part of the body exposed and the lower part submerged. Their compound eyes are split into an upper and lower half so they can simultaneously scan both the aerial and aquatic environments.

Like all animals, insects are dependent on the information they receive about their environment through their sense organs. The ability to detect sound is developed in many insects, and the sense of hearing is best developed in insects that produce sound. Many species use hair sensilla to detect sound, but the most efficient auditory organs are the membranous tympanal organs found in certain moths and grasshoppers that are sensitive to frequencies extending well into the ultrasonic range. Sound may be used to communicate between individuals, locate prey, or even avoid becoming a meal. Many owlet moths use their tympanic organs to detect the ultrasonic emissions used by echo-locating predatory bats in order to take evasive action. Oddly a parasitic mite infests the hearing organs of these moths. A tympanic organ is located on each side of the moths body but only one side is infested. This allows both moth and mites to survive. Of course this game can be played both ways. Bats use the night songs of katydids to home in on their prey.

Insects can also produce a bizarre variety of sounds by rubbing one body part against another, by the vibration of special membranes, by striking some body part against the substrate, by forcibly ejecting air or liquid from some body opening, by vibrating their wings, or by any number of other methods. Since insect activity is dependent upon temperature, it's not surprising to find that the rate of a cricket's chirp increases and decreases with temperature. Based upon the number of chirps of the common field cricket, a formula has been worked out to determine the temperature on the Fahrenheit scale. Count the number of chirps in fifteen seconds and add thirty-eight. Try this the next time you are enjoying a summer evening in your backyard.

Light is perceived in some unusual ways. The best developed and most complex light receptors in insects are the compound eyes, which are composed of many individual facets. Each lens works independently and carries its own image to the brain. The dragonfly has over 30,000 facets in each compound eye, producing a weird mosaic image of its surroundings much as if a scene were viewed through the ends of a bundle of soda straws. Since there is no means of focusing the lens, insects are shortsighted, making the compound eye well adapted only for catching quick movements rather than forming a sharp image. Visual acuity is probably only about 1/100 to 1/1000 of human vision. Some insects can perceive color but perception doesn't extend over the whole range visible to man.

Insects not only receive light but emit it as well. Bioluminescence occurs in many different types of organisms. It is best developed among insects in the beetle family Lampyridae (fireflies). Stored in the tip of the beetles abdomen are a pigment called luciferin and the enzyme luciferase. When oxygen and luciferase are combined with energy in the form of ATP (adenosine triphosphate), a catalytic reaction causes the luciferin to produce cold light. Nearly 100% of the energy released in this reaction is light. By contrast, 95% of the energy of an incandescent light bulb is dissipated as heat. Man would do well to emulate this efficiency in his own lighting devices. During the construction of the Panama Canal, it is reputed that a surgical operation was continued one night, despite a power failure, by the light of a jar of cucujos, a click beetle with light organs on its thorax. The light signals emitted by fireflies primarily serve as a means of bringing the sexes together.

"Each species with luminescent organs has a characteristic flash code, which involves the length and rate of the flashes, and a specific period of delay before a flash is answered. When a male is seeking a mate, it usually flies over an area and flashes. Stationary females located in the vegetation respond by flashing at a specific interval after the male flashes, and a male receiving a response will fly to the female. Light is the only cue involved in recognition. Neither sex recognizes the other without appropriate flashing. These insects can be attracted to artificial lights if the lights are flashed at the proper rhythm." (Borror, *et al.* 1981) In some species, the female mimics the flash response of the females of other species, attracting males of these species and then eating them when they arrive. Talk about femmes fatales! In the case of some fireflies that occur in Southeast Asia, large numbers of males gather in trees and flash in unison, lighting the whole tree. Too bad these don't come out at Christmas time.

The senses of taste and smell are the most highly developed in insects and are of great importance. The world of most insects is largely one of odor. Although the organs of taste are located principally on the mouthparts, some insects have taste organs on the antennae or feet. The sensitivity of chemoreceptors to some substances is very high, allowing detection of odors at very low concentrations and miles from their source. Chemical stimuli play important roles in a variety of behaviors. Insects secrete chemicals called pheromones which are used to communicate between members of the same species. Insects invented perfume long before man thought of the idea. Females of many species use chemicals to attract members of the opposite sex.

Insects do more than just communicate with chemicals. They are literally rambling chemical factories and many produce noxious brews to defend themselves. A well-studied insectan expert in chemical warfare is the bombardier beetle. When disturbed, this aptly named critter discharges a noxious cloud of gas as a result of the mixing of two chemical substances from separate glands. Hydrogen peroxide and hydroquinones are mixed in a reservoir with a catalytic enzyme. The combined substances reach the temperature of boiling water providing enough heat to vaporize a sizable portion and expel an irritating cloud of quinones. This defensive aerosol is ejected in quick pulses (about 500 pulses per second) rather than a continuous stream. The ejection system is similar to the pulse jet propulsion mechanism in the German V-1 buzz bomb of World War II. The beetle can fire repeatedly before having to reload, and aided by a flexible nozzle at the tip of the abdomen it makes an excellent marksman. Some of these defensive secretions are no laughing matter. One group of African grasshoppers releases secretions so toxic they can kill small mammals. Many native tribes have used a variety of insects to prepare their poison arrows for hunting.

There are many types of organs that are sensitive to mechanical stimuli such as touch, pressure, or vibration. The cockroach possesses two posterior sensory appendages called cerci from which tiny hairs project. These hairs are extremely sensitive to air movements caused by any rapid approach and trigger the escape response. That reaction time has been clocked at 11 milliseconds. That is ten times faster than the human eye blink, which takes a ponderous one-tenth of a second. By the time your newspaper slams down on the kitchen counter, the cockroach has ducked safely into its harborage, no doubt with a smirk on its face.

Many insects have startling physical capacities compared to our own. A man can pull 0.86 of his own dead weight and a horse ranges from 0.5 to 0.83. The average insect can pull up to 50 times its own weight due to an exceedingly complex muscular system and the superior leverage that can be attained by muscle attachment to an exoskeleton. Man has some 792 distinct muscles, while a caterpillar may have over 4,000 muscles. Beetles fitted with a special harness in the laboratory have been able to lift over 800 times their own weight, which is comparable to a man lifting 60 tons. And when it comes to jumping, many insects put the best Olympic athletes to shame. Fleas are known to leap for a distance of 13 inches and to a height of 8 inches. An equivalent leap for a man would carry him 800 ft horizontally and over a 30 story building. Grasshoppers are also great leapers, and resilin (the elastic protein in their legs) is the most nearly perfect elastic known, having an extraordinary 97% efficiency in returning stored energy. If a ball could be made of resilin, it would bounce nearly as high as you threw it.

The click beetles of the family Elateridae are fairly common insects. These beetles are peculiar in being able to click and jump. The clickings are made possible by the flexible union of the first segment of the thorax or chest area and a spine that fits into a groove on the lower surface of the thorax. If one of these beetles is placed on its back on a smooth surface, it is usually unable to right itself by means of its legs. It bends its thorax backwards, so that only the extremities of the body are touching the surface on which it rests. Then with a sudden jerk and clicking sound, the body is straightened out. This movement snaps the spine into its groove and throws the insect into the air, spinning end over end. It continues this action until it lands right side up. This jack-knifing motion generates an amazing 400 gravity forces (Gs) and the beetle must endure a peak brain deceleration of 2300 Gs during the movement.

Insects are the only invertebrates with wings. The ability to fly is responsible in large part for the phenomenal success of insects over the course of 400 million years. Wing strokes exceeding 1000 per second have been recorded and flight speeds of 50 MPH can be achieved. That may not seem very fast when compared with many birds or man-made vehicles but in terms of body lengths per second it may be quite remarkable. "A man running at top speed covers about 5 body lengths per second, a horse about 6, a cheetah (probably the fastest land animal) about 18, a peregrine falcon (doing 180 miles (290 km) per hour in a power dive) about 175, and a jet plane flying at three times the speed of sound about 100. A blow fly 10 mm in length flying at a speed of 3 meters per second (a normal flight speed for such a fly, and equivalent to about 6.8 miles (11 km) per hour) covers 300 times its body length per second." (Borror, et al. 1981) Dragonflies are the master aerialists with abilities to steer accurately and quickly, hover and go sideways or backward. Only the hummingbirds can approach them in ability to maneuver on the wing. Insect flight is dramatically illustrated by the migratory locust, which has devastated crops since biblical times. These pests can travel enormous distances. Clouds of them have been seen over the Atlantic up to 1200 miles from land. Swarms of billions of locusts may cover areas of several hundred square miles and take several days to pass a given point. A 3,000 sq. mile swarm of locusts hit Oregon and California in 1949.

Migrations of birds are well known to the general public, much less so are the equally incredible migratory journeys of butterflies. Few would think that such dainty creatures with such a slow fluttering flight would be capable of travelling long distances. Vast numbers of the monarch butterfly wing southward from southern Canada and the northern United States to Florida, the Gulf states, California and Mexico. From these wintering sites the mated females head north again each spring, arriving where young milkweed plants are available for the larval stages. The first brood to mature heads farther north to establish another generation. One monarch tagged on the north shore of Lake Ontario on September 18 was recovered on January 25 at San Luis Potosi, Mexico, a distance of 1,870 miles. Not bad for such a delicate creature, even with the assistance of a strong tailwind.

Few natural creations can rival the wings of butterflies for beauty, but the thousands of tiny overlapping scales that cover the translucent wing membranes have great functional significance. The colorful patterns provide important visual clues during courtship and even help protect these insects from avian predators by forming frightening eye spots. Another important function of wing scales is temperature regulation. Scales act as miniature solar heat collectors by absorbing the sun's heat and conducting it toward the body. Scales also strengthen the wing and improve its aerodynamics. Removing the scales can decrease wing lift by as much as 20 percent. Scales can also function as toxic waste dumps by storing chemicals from plants consumed by the caterpillars. These toxic compounds harmlessly placed in the wing scales can make a butterfly unpalatable to predators. When caught in a spider web, the scales may detach, allowing their owner to escape the spider's banquet table.

Of all the amazing feats of insects none is more remarkable than the change in form that accompanies development. The transformation of the homely caterpillar into the beautiful butterfly is as great a magic act as nature performs. During the pupal stage the larva nearly dissolves into a cellular soup from which the tissues of the adult are assembled. An equivalent transformation would be the change of a snake into an eagle.

Ants are everywhere but are only occasionally observed. Few appreciate their diversity, abundance and complex social organizations. Recent measurements suggest that about 1/3 of the entire animal biomass of the Amazon rain forest is composed of ants. Ants may be responsible for the formic acid found in previously unexplained quantities in the atmosphere above the Amazon forest. It is estimated that formicine ants release 10^{12} grams of formic acid globally each year. Their fascinating life cycles could entertain for several lifetimes. No doubt this is the reason entomologists would like to believe in reincarnation. Some entomologists are so fascinated with these creatures that they distinguish themselves with the title myrmecologist. Even the ancients were impressed by the ant's industry, as evidenced by the biblical injunction "go to the ant thou sluggard, consider her ways and be wise". However scientific observations have demonstrated that like human beings, ants display striking differences in temperament. Contrary to common belief not all ants are industrious. In fact, some don't work at all except under extreme pressure. These shirkers just hang around watching others or run about as if they really were busy. Solomon might never have uttered his admonition had he observed one of the many species of slave making ants. Periodically, slave-makers raid colonies of their own or closely related species. Usually only the eggs, larvae and pupae are kidnapped. The eggs and larvae are generally eaten but worker pupae are allowed to mature and are used as slaves. Captured ants fully accept their role and become dedicated slaves, feeding and caring for their captors and doing all sorts of menial chores. The social attachment of slaves to slave-makers is so complete that they even accompany the parasites on raids against colonies of their own species. Many North American species of slave-makers live in acorns that have been hollowed out by beetle larvae. A colony of about 15 slave-makers requires about 50 slaves. Slave-makers are totally dependent on their slaves and would starve to death without them even if food were nearby. No doubt many American house-wives have noticed a social similarity with the institution of marriage.

An interesting mutualistic association exits between ants and aphids. No doubt many people have watched ants and aphids on the same plant without realizing the significance of the association. Aphids often excrete a sugary substance called honeydew, which is partially digested plant sap. Large groups of aphids are tended by ants in a manner that has been compared by some naturalists to cattle farming. In return for the honeydew, which is produced on demand in response to stroking by the ant, the aphids receive protection from predators, are moved from plant to plant, and are provided other services. Some ants transport aphids or their eggs from their host plant in the fall to underground chambers where they are cared for over the winter. In the spring the ants return the aphids to their host plant.

Other interesting ants belong to a group known as honey ants because some of the workers function as honey-pots or reservoirs for storing honeydew gathered by other worker ants. The workers that store the honey are known as rotunds or repletes because of their grossly distended abdomens, which may reach the size of a small grape. They act as living barrels of stored food and hang from the ceiling of special chambers. During the winter, the rotunds regurgi-tate food on demand by normal workers. In some areas of the world, natives have made a fermented beverage by collecting and pressing the rotund honeydew ants. Man has found many other unusual uses for ants. In Texas, Central and South America, the soldiers of leaf cutting ants of the genus *Atta* have huge saw-toothed jaws capable of a powerful grip that is maintained even when ants are beheaded. Natives along the Amazon have used these ants to suture deep wounds with the primitive method of "clamps stitching."

The nests of ants provide a home for an astonishing variety of insects. Several hundred different species have been recorded from ant nests throughout the world and many have complex relationships with their hosts. Chemical attraction or deception is frequently used by invaders that may spend much of their life cycle inside the nest. Normally only colony members are allowed to enter the ant society and foreign intruders are dealt with harshly. By using various techniques, a considerable number of solitary arthropods have managed to penetrate ant nests, suggesting that these guests have broken the ants' communication and recognition codes. Once inside the nest they may function as scavengers, parasites, or predators and become dependent on the ants during part or all of their life cycle. Rove beetle larvae are readily accepted, groomed, and fed by their ant hosts as a result of a chemical attractant secreted by a row of paired glands along the sides of the beetle abdomen. Ungrateful, the beetle larvae eat small ant larvae in return.

Insects make their living in unimaginably strange ways. Life cycles are varied to fit unique niches in nature's web of life. Among the thousands of species of parasitic wasps, life cycles are often highly specialized. The adult females of certain species enter water to seek out aquatic insect hosts to parasitize. They move through the water by flapping their wings similar to a method used by swimming penguins. *Rhyssa*, one of the largest parasitic wasps seeks out the larva of the giant wood-wasp, which bores through conifer wood. *Rhyssa* has a thread-like ovipositor 1 1/2 inches long — longer than her head and body — with which she drills through the bark and solid wood to lay her eggs on her victim deep in its tunnel. Remarkably, *Rhyssa* detects her victim inerrantly and manages to insert an ovipositor no wider than a human hair deep into solid wood and then squeeze an egg through it.

Galls are abnormal growths of plant tissue caused by a stimulus of some sort. There are hundreds of species of tiny gall wasps that induce the formation of galls by insertion of their eggs into plant tissue. Many gall wasps have a complex life history and form a characteristic gall on a particular part of a particular plant. *Rhodites* causes a gall known as robin's pincushion. Another species of gall wasp waits until this gall has formed and then lays its own eggs in it much like a cowbird unable to build a nest of its own lays its eggs in the nests of other birds. Both species are then subject to the attack of a variety of insect parasites. The parasitic wasp *Torymus* uses her long ovipositor to pierce the gall to lay her eggs on the *Rhodites* larvae. Using the same method another tiny wasp, *Oligosthenus*, specializes on the larvae of *Rhodites'* nestmates. Neither parasite makes the mistake of laying its eggs on the wrong host.

The larvae of many flies are parasitic on animals, but some take a roundabout way of finding their host. The larvae of the human bot fly invade the tissues of man and cause boil-like eruptions. The female fly, instead of laying eggs directly on the human host, captures a mosquito or other biting insect and, holding it firmly against her underside, glues a bunch of eggs to the body of the captured insect. After release, the eggs of the parasite are transported to the human host by the parasitized mosquito as it goes in search of a blood meal. The warmth of human skin stimulates the bot fly larvae to emerge from their eggs and burrow into the skin, where they will live for the next 1 to 2 months.

Insects are the masters of camouflage. They imitate environmental objects such as twigs, leaves, thorns and even bird droppings. Many insects resemble leaves, complete with blemishes and fungal spots in addition to the proper leaf tint and venation. They orient themselves on the stem in the appropriate manner and flutter in the breeze as convincingly as any real leaf. Certain Malaysian preying mantids greatly resemble flowers. Unsuspecting prey attracted visually by these "blossoms" become a tasty meal. Insects also copy each other. Mimicry is usually visual but smell, sound or behavior may also be duplicated. Ant mimics run alertly over the ground in the style of true ants. A palatable species of insect can avoid being eaten by predators by imitating the appearance of an unpalatable, venomous, or otherwise undesirable species. This type of mimicry is quite common, especially in butterflies. The viceroy butterfly looks much like its distasteful model the monarch. Many flies look like bees and wasps, though they lack the fearsome stings to defend themselves. Such flies may add to their deception by duplicating the sound of their model. The buzz of a wasp wing is around 150 hz and its drone fly mimic a nearly identical 147 hz. Even more remarkable, an insect may imitate entirely different models as it changes in size and shape during growth and metamorphosis.

Insects have no conscience when devising feeding strategies. Certain assassin bugs have forelegs covered with secretory hairs that exude small droplets of a highly viscous substance which looks like dew. Flies attracted to the dew are entrapped and quickly consumed. The Javan bug, *Ptilocerus*, has a tuft of bright red hairs on its body, marking the spot where a gland opens beneath the abdomen. Secretions from this gland are very attractive to ants. However, the secretion apparently has a narcotic action once it is eaten and the groggy ant is pierced and rapidly sucked dry by the bug.

Insects make a fiction writer's job easy. No vivid imagination is needed. No creature too bizarre can be conjured up by the writer's mind. All one has to do is observe the great natural drama going on around you. Nature is a public library and few tomes more fascinating can be found. What's more, the last book hasn't been written yet. Evolution is an unfinished story. So slow down from life's harried pace and read a few tales. Insects and their varied ways of life can be an endless source of wonder and delight. You certainly won't regret the time spent.

"For every man the world is as fresh as it was at the first day, and as full of untold novelties for him who has the eyes to see them."
                                                                T. H. Huxley

References*

Borror, D.J., D.M. DeLong, and C.A. Triplehorn. 1981. An Introduction to the Study of Insects. Saunders College Publishing. 827 p.

Evans, H.E. 1978. Life on a Little Known Planet. E.P. Dutton 324 p.

* Permission of the publishers is granted to quote from these books as indicated in the text.

Entomophobia:
The Case
for Miss Muffet

Little Miss Muffet Sat on a tuffet, Eating of curds and whey; There came a
big spider, And sat down beside her, And frightened Miss Muffet away.
(Mother Goose 1916)

Tad N. Hardy

ENTOMOPHOBIA, the irrational or unreasonable fear of insects, and the related arachnophobia, fear of spiders, are only two of
more than one hundred terms describing recognized human fears
(Terhune 1949). Among the common fears reported by Western
societies that involve animals, insects and spiders rank relatively high
as fear-inducing organisms, running close behind rats and snakes.
This ranking may be somewhat surprising from an entomological
perspective, although many entomologists would concede that public
sentiment rarely favors insects. What is it about insects and spiders
that evokes such negative responses from a significant portion of the
population? Why are they feared? And is there a little Miss Muffet
in all of us?

Insects and their activities have received mixed reviews with
respect to societal attitudes. Their abrupt appearance in a household
dwelling, for example, has been considered both a good omen and an
ominous indicator of impending disaster in traditional folklore (the
chirp of a cricket versus the click of the deathwatch beetle). Most
fables and nursery rhymes, in contrast to the one above, portray
certain insect groups favorably as hard workers (bees and ants), as
merry minstrels (crickets and katydids) or as symbols of beauty
(butterflies). Such attributes are seldom ascribed to snakes or rats in
Western juvenile literature, apparently reinforcing societal perceptions of these animals as dangerous and ill-tempered. These stories
may often represent the extent of a young child's exposure to insects
except for occasional home encounters or very limited field experience. Older children and adults, via informative books and television
programming, also have opportunities to view insects and their life
histories in an interesting and positive way, although less benign
presentations are common in literary fiction or movie theaters
(Mertins 1986). Yet amid a prevailing tendency to represent insects
favorably, fears persist in children and adults alike.

The anxiety produced in a fearful individual by a potential encounter with an insect ranges from mild aversion to (albeit rare) stark terror. The degree of negative response to insect- or spider-related stimuli is important in assessing the severity of an entomophobic problem, if one exists. It is the magnitude of this problem and its subsequent consequences that distinguish between common fear and true phobia. Common fear is a natural extension of human experience. It is nearly universal in nature and represents a reasonable, appropriate response to situations that involve potential danger or require caution. Phobias, however (from the Greek *phobos*, meaning panic-fear or terror), are characterized by persistent, disproportionate anxiety in situations of low inherent harm (Rachman 1968, Marks 1969, Beck *et al.* 1985) and thus are quite restrictive. Many individuals may express a fear of insects or spiders, but few are considered truly phobic because their ability to function in a normal daily routine is not impaired by their fear (Marks 1969, Agras 1985).

Phobia

Intense fear of insects or spiders is classified as a simple phobia and as such is grouped with other animal phobias, fear of heights, fear of closed spaces, and so on. Most phobic individuals recognize that their fear is unfounded, yet they involuntarily respond in a fearful manner to the thought or presence of a spider or insect or situations associated with them:

> As with any phobia, [insect and spider phobias are] marked by an avoidance of places where the feared objects are likely to be encountered, in the case of spiders, cellars and dark cupboards would be among the places avoided. The spider phobic may live in a protected environment well sprayed with insecticide, or policed by a friend or relative. Before retiring for the night, the phobic may fearfully search the room for the insect, probably well armed with a can of insecticide. Or, better still, someone else will search the bedroom and declare it safe. Nightmares about spiders are common, and the sight of one leads to immediate flight. (Agras 1985)

General symptoms associated with simple phobias are those consistent with common fear. Sudden, intense feelings of apprehension, feelings of impending doom, fear of losing self-control, shortness of breath, increased heart rate, faintness, trembling, and sweaty palms are typically reported (Rachman 1968, American Psychiatric Association 1980, Goodwin 1983). Rachman separates the phobic reaction into three components: subjective (feelings of fear), autonomic (physiological responses), and motor (immobilization or flight). Occasionally these components are expressed with striking clarity:

> A woman with a fear of spiders screamed when she found a spider at home, ran away to find a neighbour to remove it, trembled in fear and had to keep the neighbour at her side for two hours before she could remain alone at home again; another patient with a spider phobia found herself on top of the 'fridge in the kitchen with no recollection of getting there — the fear engendered by sight of a spider had induced a brief period of dissociation. If in treatment sessions patients are brought too close to the animal, an acute panic ensues immediately in which they sweat and tremble and show all the features of terror which will wake them even from a deep hypnotic trance. (Marks 1969)

In the case of a seven-year-old boy with intense bee phobia:

> He described the subjective aspect of his reaction very simply as follows: "I am very frightened of the bees and feel scared that they will sting me and hurt me." His mother described the autonomic reaction which her child experienced when confronted with a bee: "He used to go white, sweaty, cold and trembling, and his legs were like jelly." The child's motor reaction was a combination of flight and total avoidance . . . On at least two occasions he had run across busy streets when confronted by a bee and his parents were worried that he might come to harm because of his intense and sudden escape reactions. (Rachman 1968)

One key component inherent to the above accounts is avoidance. Insect and spider phobics actively avoid the feared object and ultimately avoid any activity that might force a confrontation with it. Unfortunately, each avoidance episode that prevents the onset of unsettling fear responses provides positive reinforcement for the entomophobic and leads to future avoidance. This vicious cycle engrains the fear rather than diminishes it (Rachman 1968, Sutherland et al. 1977).

A second component in the spread and maintenance of insect aversion is the tendency to associate the feared insect with previously neutral stimuli or to generalize the fear response to encompass whole groups of organisms or objects (Rachman 1968, Bennett-Levy & Martean 1984, Beck et al. 1985). An arachnophobic may fear cobwebs, then a cupboard with cobwebs, and finally the cupboard itself. Intense fear of wasps may be generalized to include flies, moths, or even machinery that makes a buzzing sound.

Although many in the general population prudently respect the potential threat posed by certain insects (such as wasps) or even possess a healthy fear of them, insect phobics do not require a logical basis for their fear. Thus theories of how and why this phobia develops and persists are of interest.

Origin, Onset, and Incidence

The etiology of insect and spider phobias has been and continues to be a topic of considerable debate (Sturgis and Scott 1984, Suomi 1986). For centuries it was believed that expectant mothers frightened by an insect (or any other experience) passed their fear to the developing child. Thus the child of an insect-startled mother was destined or "marked" to be insect-shy. In the early 1900s Sigmund Freud applied his psychoanalytic theory of repressed sexual and aggressive childhood drives to a simple animal phobia (a horse phobia, [Freud 1909]). He interpreted the excessive anxiety displayed by a five-year-old boy to have resulted from displacement of jealousy and hostile attitudes toward his father to an avoidable object (in this case, an animal). Typical phobic symptoms of fear and avoidance persisted, according to psychodynamic theory, because the animal symbolized the feared retaliation of the boy's father for his repressed thoughts. Subsequent psychoanalysts pronounced all phobias (including those of insects and spiders) concrete representations of unconscious mental conflicts (Marks 1969).

A somewhat bizarre extension of the psychodynamic theory to arachnophobics has been proposed by Adams (1981). He concluded that a fear of spiders can be traced to a tendency to equate the predatory spider with female love, sexual desire, and femininity. The phobia reflects childhood inability to separate mother from self and is tied to our urban, capitalistic society. This complex interpretation relies heavily on symbolism, a crucial psychoanalytic component; however, symbolism as a causative factor of simple insect or spider phobias is difficult to substantiate.

The prepotency theory (Marks 1969) and the similar preparedness theory (Seligman 1971) of simple phobias suggest that insect and spider phobias are derived from innate common fears of naturally occurring and potentially harmful objects. This biological viewpoint states that fear of these organisms conferred a selective survival advantage to pretechnological humans and that phobic tendencies develop from these remnants of adaptive fears. A substantial amount of observational and experimental data exist to support this idea (Sturgis and Scott 1984). In addition, a fear of insects and spiders (and other animals in general) universally and spontaneously develops in early childhood and then diminishes with age. Although support for the theory that phobias are residual biological implants ranges from ardent (Agras 1985) to somewhat skeptical (Sutherland *et al.* 1977), many psychiatrists and therapists agree that it may explain partially the apparent nature-nurture etiology of entomophobia. There is even discussion that some individuals are physiologically more susceptible to phobic symptoms and that susceptibility may be hereditary. Again, separation of innate and learned influence is difficult.

A recent study by Bennett-Levy and Martean (1984) offers an alternative hypothesis to preparedness theory. They suggest that fears of insects and spiders may not be related to these organisms' biological significance but rather are based on their "fear-evoking perceptual properties." Their study examined questionnaire responses of adult subjects to perceived characteristics of selected animals (Table 1.) Participants reported ugliness traits to include compactness of body, antennae projecting from body, and an odd relation of eyes to head. Other visual and tactile cues also negatively influenced perception (spiders have long, hairy legs and would feel unpleasant; cockroaches move very fast). Ugliness, sliminess, and speed of movement were positively correlated with fear response and a desire to avoid the insect (nearness rating). Bennett-Levy and Martean concluded that animal characteristics that are divergent from human form evoke aversion that leads to fear.

Behaviorists postulate altogether different origins for insect and spider phobias. Their theories favor the fear of insects as a conditioned, or learned, response to certain stimuli (the nurture side of a nature-nurture etiology). Entomophobic symptoms develop when the innate capacity to fear is channelled toward insects because of a previous negative experience or possibly because other individuals have been observed responding in a fearful manner to insects. A traumatic insect phobia might be triggered by a single event, such as a multiple wasp sting or a cockroach up the pant leg. For insect or spider phobias that develop in adulthood, the phobic individual often may recall such an event vividly (Beck *et al.* 1985).

It is unlikely, however, that all entomophobics have had such a traumatic experience. Vicarious acquisition of a fear of insects through observation or repeated oral warnings from parents or peers also is theorized to reinforce a developing aversion (Beck *et al.* 1985), although some disagree (Goodwin 1983). Experimental evidence has shown that specific fears can be transmitted in this manner (Bandura and Rosenthal 1966) and may run in families, particularly from mother to daughter (Marks 1969).

Clearly, no single theory accounts for fear of insects and spiders. It is likely that both learning and experience modify or direct existing biological tendencies to fear these organisms.

Table 1. Mean ratings for fear, nearness, and animal characteristics.

Animal	Ugly	Slimy	Speedy	Moves Suddenly	Fear[a]	Nearness[b]
Spider	2.43	1.06	2.25	2.52	1.64	2.88
Beetle	2.10	1.18	1.55	1.57	1.33	2.50
Moth	1.53	1.09	2.04	2.32	1.25	2.27
Ant	1.86	1.04	2.04	2.14	1.14	2.22
Gresshopper	1.76	1.12	2.48	2.77	1.16	2.06
Caterpillar	1.65	1.24	1.14	1.12	1.05	1.84
Butterfly	1.06	1.02	2.08	2.36	1.00	1.33
Rat	2.24	1.10	2.35	2.53	2.08	3.90
Grass Snake	1.80	1.78	2.12	2.42	1.55	2.78

Rat and snake included for comparison. Adapted from Bennett-Levy and Martean 1984. $n= 113$.
[a]Characteristics and fear scale 1, not afraid; 2, quite afraid; 3, very afraid.
[b]Nearness scale: 1, enjoy picking up; 2, unpleasant, but would pick up; 3, touch or within 6 in.; 4, stand 1-6 ft. away; 5, move >6 ft. away.

There is no question that the onset of entomophobia and arachnophobia almost always occurs in childhood, when a generalized fear of animals emerges (Marks and Gelder 1966). Agras *et al.* (1969) stated that common fears of animals develop during childhood or possibly during early adulthood. Marks (1969) placed the onset between one and seven years of age, and Goodwin (1983) states that animal phobias usually appear at two to four years of age. The fear diminishes with time, and if an insect or spider phobia is not apparent by age ten or twelve, it probably will not occur (although Goodwin states that a fear of spiders is more common among adults than it is among children). It is within this five-to-seven-year developmental window that children are most susceptible to converting a normal fear response to insects into a full-blown phobia. Most outgrow the fear, but a select few carry it with them into adulthood. Beck *et al.* (1985) suggest that an unpleasant experience (actual or vicarious) during this period of development may "fixate" the fear, triggering the onset of a simple phobia. Entomophobia can begin in adulthood, but cases are uncommon. It is interesting that fears of insects and spiders acquired in adulthood are the most difficult to treat or eliminate (Agras 1985).

The prevalence of insect and spider phobia in the general population is difficult to estimate. Treatment statistics comprise the primary source of frequency data but are biased because they reflect only individuals compelled to seek clinical treatment because of their intense fear. Thus, clinical data may underestimate the proportion of single-animal phobics in the nonclinical population (Agras *et al.* 1969, Goodwin 1983). A much higher proportion of the population considers itself squeamish with regard to insects, and some express outright fear albeit to a degree not compatible with true phobia.

Figure 1 represents a theoretical relative distribution of attitudes toward insects and spiders in the general adult population based on general statements from recent texts addressing phobias. Marks (1969) estimated the incidence of true phobic response to animals to be relatively rare (3%). A survey study reported by Agras (1985) estimated that only 1.1% of the population exhibits specific animal phobias, many of which include insects or spiders. The Epidemiologic Catchment Area Program, a comprehensive survey of psychiatric disorders conducted by the National Institute of Mental Health, reports the incidence of all adult simple phobias at about 10%, but insect, spider, and other animal phobias constitute only a portion of this figure (Robins *et al.* 1984).

Two important factors to consider when estimating frequencies of insect phobics are age and gender. A generalized fear of animals is more common among children than among adults. During childhood, boys and girls appear to have an equal probability of developing an insect or other animal phobia (Marks 1969). Fear and phobic persistence following puberty, however, is much greater among women than among men (Marks and Gilder 1966, Marks 1969, Abe 1972). Geer's (1965) Fear-Survey Schedule compared mean scores of male and female reactions to 51 items using a seven-point selfevaluation scale (1, no fear to 7, terror). Female response was higher for stinging insects (M = 1.80, F = 2.38) and significantly higher for spiders (M = 1.20, F = 2.19, P < 0.01). Marks (1969) found the overall incidence of animal phobia to be 9:1 for women to men. Nearly twice as many women as men reported feelings of fear (Agras 1985), particularly toward insects, spiders, and other animals. Examination of clinical studies, case histories, and volunteers for behavioral experiments reveals that the majority of entomophobic patients and volunteers for insect-spider fear studies are women.

Several explanations could account for the disproportionate number of female entomophobics. Men may be less willing to admit to a fear of insects and spiders in surveys and interviews, and they may hesitate to seek treatment for their fear. A marked decrease in

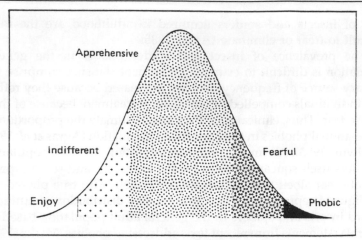

Figure 1. Estimated relative frequency distribution of attitudes toward insects and spiders in the general population.

expressed fears by males at puberty supports this explanation. Social conditioning also may be important, such as different treatment of children expressing a fear of insects depending on the child's sex. A classic study by Lapouse and Monk (1959) suggests that social conditioning concerning insects may begin during the onset period of natural fears before a hesitancy to admit these fears has developed. Their sample was drawn from 482 randomly selected households containing 6 to 12 year olds, in which mothers and children were interviewed. Based on the mothers' responses, 12% of the boys and 40% of the girls (x = 26% for all children) were determined to have a fear of "bugs." Fifty-six percent of the children agreed with their mothers' assessment, and an additional 27% expressed a fear that was not evident in the mothers' responses.

Bowd (1984) examined the correlation between pet ownership and fear of animals (including spiders) in children and concluded that pet care responsibilities may reduce the fear of animals. In a survey of female college freshmen, however, previous experience with or knowledge of animals was not conducive to reducing fears of spiders. Bowd and Boylan (1984) found that 53% of freshmen completing high school biology expressed a fear of spiders compared with only 35% of those who had not taken such a course. These data suggest that social conditioning in childhood may strongly affect subsequent development of entomophobia and that casual conditioning in adulthood is not an effective determinant. One-on-one interactions of children with peers or respected adults seem to have a much greater influence on conditioning than does exposure through nursery rhymes and fables.

Treatment

It is not surprising that fear reduction therapy for entomophobic and arachnophobic individuals is as varied as the theories that seek to explain these phobias. Basically, there are two divergent schools of thought regarding treatment of insect and spider phobics. The first is traditional Freudian psychoanalysis, which attempts to bring repressed, underlying causative factors to the surface for use in complex analysis of psyche development gone awry (recall the theory of spider phobia origin that equates a fear of spiders with mother-self, male-female conflict). Psychoanalytic therapy for entomophobics has a limited number of supporters today, but controlled clinical studies are lacking. The second approach to fear reduction, particularly in children, is behavior therapy. Support and enthusiasm for this branch of therapy has grown steadily since the 1950s as observational and empirical evidence for its successes has accumulated. Today, behavior therapy is the recommended course of action for treating simple insect and spider phobias, and it enjoys a reasonably high success rate.

Most current behavior therapy methods used for treating simple phobias are variations on a basic thesis set forth by Wolpe (1958), who proposed that phobias are a learned condition and, as such, can be unlearned. His contention was a unique application of classical behavioral conditioning theory to insect and other simple phobias. Although widely accepted in principle, Wolpe's prescribed treatment procedures have been expanded and modified to some extent.

Even among behavior therapists, there is no consensus about which method is best for treating subjects with a fear of insects or spiders. The choice of therapy depends on the age of the fearful

---

Methods of behavior therapy for treatment of insect and spider phobias*

    I. Desensitization
        A. Basic systematic (gradual imagery)
        B. *In vivo* (gradual exposure)
    II. Implosion (intense imagery)
    III. Flooding (intense *in vivo* exposure)
    IV. Contingency (positive reinforcement)
    V. Modeling (vicarious exposure)

* Adapted from Sturgis and Scott (1984), Morris and Kratochwill (1985).

individual, the extent of the problem, and the experiential bias of the therapist. Most insect phobics who seek treatment, however, will begin a program of some form of desensitization.

Systematic desensitization (Wolpe 1958) is an imagery-based approach that is probably the best studied of all behavioral treatments. It usually involves preliminary muscle relaxation followed by visualization of increasingly threatening images of the feared entity Obviously this procedure can become laborious and may require a number of therapy sessions depending on each session's success. Several variants of this method have arisen that suggest that relaxation may not be necessary (Marks 1978, Agras 1985) and that actual exposure to an insect or spider may be more effective.

In vivo desensitization techniques rely on increasing physical exposure to the feared insect over time to reduce fear responses. Often, systematic and in vivo therapies are intertwined through the use of imagery aids such as photographs or drawings. Sarafino (1986) outlines a typical anxiety reduction regimen used to treat a five-year-old with a fear of flying insects.

Implosive therapy also relies on imagery therapy but, unlike desensitization, there is no gradual mental exposure. Instead, the insect or spider phobic is asked to visualize a confrontation with the feared insect that ends in catastrophe. "Imagine you are a fly, trapped in the sticky web of an enormous spider . . ." and so on. Fazio (1970, 1972) examined the effectiveness of implosive therapy in several studies using adult females who exhibited semiclinical insect phobias. He found no significant differences in various measures of fear reduction between implosion and control therapies consisting only of supportive statements of encouragement or noninsect-related imagery Rachman (1966) reported basic desensitization to be a superior treatment to limited implosion therapy for spider phobics. However, these early implosion methodologies have been criticized for using sessions of too short a duration to be effective (Mavissakalian and Barlow 1981). Current implosion therapy is administered for extended periods to allow the anxiety to habituate or disappear.

Flooding therapy is similar to implosion therapy but invokes actual exposure to the feared insect rather than relying on the phobic's imagination. In theory, an extreme flooding session might include strapping the cockroach phobic to the floor in a dark room containing hundreds of cockroaches. In practice, of course, such sessions do not occur. Instead, fearful individuals are exposed to a cockroach (the larger the better) in a reasonable manner for a single period of several hours, usually with the aid of a therapist, until the phobic's anxiety begins to subside. Flooding is not the treatment of choice for many therapists (or their patients, for that matter), but ample evidence exists to support its success (Sturgis and Scott 1984).

One self-help book strongly advocates this method and details procedures for self-treatment (Sutherland *et al.* 1977).

Additional varieties of behavior therapy have been used to a lesser extent as treatments for insect and spider phobias, either alone or in conjunction with other methods. Of these, modeling has enjoyed some degree of successful application. Modeling involves approach toward and contact with the feared insect by a therapist, parent, or respected peer in the presence of the phobic individual. This serves to alleviate the phobic's initial perceptions of fear and is designed to result in imitative behavior by the phobic. Positive reinforcement that incorporates a merit or reward system may be a particularly useful supplement to therapy with children. Edwards (1978) reported the case of "Little Tom," a four-year-old insect-spider phobic, who was encouraged to start an insect collection as part of a regular desensitization therapy He received a desired reward with each collection and soon had acquired quite a different perspective of insects. Two years later he was still collecting.

Drug therapies for physiologically reducing anxiety have been used sparingly in modern treatments for adult phobics. Mild tranquilizers or beta blockers may alleviate symptoms; in isolated cases, stronger tranquilizers may be enlisted, particularly when phobic behavior poses a threat to the well-being of the individual. But the nature of insect or spider phobias and the tendency to control them using behavioral methods argue against consideration of drug therapy as a viable method of treatment (Goodwin 1983, Agras 1985).

The goal of treatment therapies is to reduce the anxiety of the entomophobic to a point at which the avoidance behavior that previously directed the phobic's actions no longer interferes with normal activities. Current information suggests that this goal is best accomplished by some form of exposure to insects or spiders combined with ample moral support (Andrews 1966, Rachman 1968, Marks 1969, Sutherland et al. 1977, Agras 1985).

Common sense and a little understanding are extremely important in assisting or perhaps preventing fears of insects and spiders. The irrational nature of entomophobia in adults increases the difficulty of calming the fear through education about insects or explanations of their activities, although some individuals may be helped by such knowledge. Children, however, may greatly benefit from active, positive instruction about insects and spiders, particularly during the developmental years when fears of these organisms become apparent. Sarafino (1986) has suggested a number of practical exercises and activities designed to prevent natural, common fears from reaching phobic proportions. Among the solutions are cultivating attitudes of tolerance toward insects among parents and other admired adults and ensuring that children see realistic portrayals of insects and spiders on television or in movies. Conversation that emphasizes positive characteristics of insects (they make honey, they pollinate flowers) is helpful, as are exposure to records or tapes of insect sounds and active use of "bibliotherapy" through insect-oriented children's books, such as *We like Bugs* by G. Conklin (1962).

Acknowledgement

The comments of A M. Hammond, J. A. Lockwood, and D. P. Pashley are greatly appreciated. Thanks are extended to Debbie Woolf for typing the manuscript. This manuscript was approved for publication by the Director of the Louisiana Agricultural Experiment Station as manuscript 87-17-1447.

References Cited

Abe, K. 1972. Phobias and nervous symptoms in childhood and maturity: persistence and associations. Br. J. Psychiatry 120: 275-283.

Adams, K. A. 1981. Arachnophobia: love American style. J. Psychoanal. Anthropol. 4: 157-197.

Agras, S., D. Sylvester & D. Oliveau. 1969. The epidemiology of common fears and phobias. Compr. Psychiatry 10:151-156.

Agras, S. 1985. Panic. W H. Freeman, New York.

American Psychiatric Association. 1980. Diagnostic and statistical manual of mental disorders (III). American Psychiatric Association, Washington, D.C.

Andrews, J. 1966. Psychotherapy of phobias. Psychol. Bull. 66: 455-480.

Bandura, A. and T. L. Rosenthal. 1966. Vicarious classical conditioning as a function of arousal level. J. Pers. Soc. Psychol. 3: 54-62.

Beck, A. T., G. Emery and R. L. Greenburg. 1985. Anxiety disorders and phobias. Basic, New York.

Bennett-Levy, J. and T. Martean. 1984. Fear of animals: what is prepared? Br. J. Psychol. 75: 37-42.

Bowd, A. D. 1984. Fears and understanding of animals in middle childhood. Genet. Psychol. 145: 143-144.

Bowd, A. D. & C. R. Boylan. 1984. Reported fears of animals among biology and non-biology students. Psychol. Rep. 54: 18.

Conklin, G. 1962. We like bugs. Holiday House, New York.

Edwards, S. S. 1978. Multimodal therapy with children: a case analysis of insect phobia. Elem. Sch. Guid. Couns. 13: 23-29.

Fazio, A. F 1970. Treatment components in implosive therapy. J. Abnorm. Psychol. 76: 211-219.

Fazio, A. F. 1972. Implosive therapy with semi-clinical phobias. J. Abnorm. Psychol. 80: 183-188.

Freud, S. 1909. A phobia in a 5-year old boy, pp. 5-149. *In* J. Strachey [ed.], The standard edition of the complete psychological works of Sigmund Freud, vol. X. Hogarth, London.

Geer, J. H. 1965. The development of a scale to measure fear. Behav. Res. Ther 3: 45-61.

Goodwin, D. W. 1983. Phobia: the facts. Oxford University, New York.

Lapouse, R. and M. A. Monk. 1959. Fears and worries in a representative sample of children. Am. J. Orthopsychiatry 29: 803-818.

Marks, I. M. 1969. Fears and phobias. Academic, New York.

Marks, I. M. 1978. Behavioral psychotherapy of adult neurosis, pp. 493-547 *In* S. Garfield & A. E. Bergin [eds.], Handbook of psychotherapy and behavior modification, 2nd ed. Chichester: John Wiley, New York.

Marks, I. M. and M. G. Gelder. 1966. Different ages of onset in varieties of phobias. Am. J. Psychiatry 123: 218-221.

Mavissakalian, M. and D. H. Barlow (eds.). 1981. Phobia: Psychological and pharmacological treatment. Guilford, New York.

Mertins, J. W. 1986. Arthropods on the screen. Bull. Entomol. Soc. Am. 32: 85-90.

Morris, R. J. and T. R Kratochwill. 1985. Behavioral treatment of children's fears and phobias: a review. School Psychol. Rev. 14: 84-93.

Mother Goose. 1916. The real mother goose. Rand McNalk Chicago.

Rachman, S. 1966. Studies in desensitization - II. Flooding. Behav. Res. Ther. 4: 1-15.

Rachman, S. 1968. Phobias. Their nature and control. Thomas, Springfield, Ill.

Robins, L. N., J. E. Helzer, M. M. Weissman, H. Orvaschel, E. Gruenberg, J. D. Burke, Jr., and D. A. Regier. 1984. Lifetime prevalence of specific psychiatric disorders in three sites. Arch. Gen. Psychiatry 41: 949-958.

Sarafino, E. P. 1986. The fears of childhood. Human Sciences, New York.

Seligman, M. E. P. 1971. Phobias and preparedness. Behav. Ther. 2: 307-320.

Sturgis, E. T. and R. Scott. 1984. Simple phobia, pp. 91-141. *In* S. M. Turner [ed.], Behavioral theories and treatment of anxiety. Plenum, New York.

Suomi, S. J. 1986. Anxiety-like disorders in young non-human primates, pp. 1-23. *In* R. Gittelman [ed.], Anxiety disorders of childhood. Guilford, New York.

Sutherland, E. A., Z. Amit and A. Weiner. 1977. Phobia free. Stein & Day, New York.

Terhune, W. B. 1949. The phobic syndrome: a study of eighty-six patients with phobic reactions. Arch. Neurol. Psychiatry 62: 162-172.

Wolpe, J. 1958. Psychotherapy by reciprocal inhibition. Stanford University, Stanford, Calif.

Used with permission from Bull. Entomol. Soc. Amer. 34(2): 64-69 (1988).

Adult Amateur Experiences in Entomology: Breaking the Stereotypes

Janice R. Matthews

In *The Pleasures of Entomology*, Howard Evans (1985, p. 27) concludes his first chapter by musing that:

> Entomologists, after all, do have to earn their keep . . . no one will pay them for enjoying themselves. The day of the amateur, defined as one who has other means of support and studies insects just for sport, is largely past.

Who makes up this supposedly endangered species called the amateur entomologist? Suppose I were to ask you to describe a type specimen. What would be this person's sex? age? personality type? How much science would the person know? What would his or her general interests and amateur activities include?

As background for an informal conference at the 1987 annual meeting of the Entomological Society of America, I began to search for answers to such questions. With the assistance of Karen Strickler, I contacted seven entomological societies whose memberships include a sizeable number of amateurs: the Cambridge Entomological Society, the Xerces Society, the Lepidopterists' Society, Butterfly Lovers International, the Amateur Entomologists' Society, the Oregon Entomological Society, and the Young Entomologists' Society (the lineal descendant of TIEG, the Teen International Entomology Group). Amateurs in these societies were encouraged to write and share their experiences. Then, to each of the 69 people on the Young Entomologists' Society mailing list who had identified themselves as adult amateurs residing in the United States, I wrote a personal letter with the following open-ended questions:

- When did you first become interested in entomology?
- How does this interest find expression in your life now?
- How does it relate, if at all, to what you do for employment?
- To what degree do you share this interest with other scientists or nonscientists?
- What other things do you feel I should be sure to mention when talking about this subject to an audience that will be made up almost

completely of professionals - and students intending to become professionals?

I received 35 letters in response, from all over the country (Matthews 1988). From the number of responses, and the length and detail of most of them, it was clear that some commonly held stereotypes need revamping.

The Adult Amateur: Beginnings of a Profile

Upon hearing the words "amateur entomologist," you probably first envision a child, for most people, a 10-year-old boy with a butterfly net. However, this study focused on those other amateurs, the adults, some of the realities of their experiences and some of the factors behind their sustained, unpaid but not unrewarded interests.

Those who responded to the survey were distributed fairly evenly across all age groups. Slightly higher numbers of individuals in the 21 to 30-year-old age group are explained by the inclusion of a few college students who currently consider themselves amateurs, whether or not they eventually plan an entomological career.

Four out of five respondents report having become interested when they were children; some remember firmly established interests when they were only four to five years old. One of every five respondents, however, remembers no particular interest in science until well into adulthood. This is, perhaps, an encouraging note, for it contradicts the idea (Miller and Prewitt 1979) that early positive science experiences are a necessary prerequisite to positive adult attitudes.

Most published accounts of amateur entomology are reminiscences written by individuals who have maintained a lifelong interest in insects. Are their experiences typical? I found it significant that not one respondent reported first becoming interested in entomology as a teenager. This age period, when peer pressures are all-important and highly contagious, appears to have been a difficult one even for highly committed amateurs. A full quarter of the respondents who had been interested as children report ignoring entomology during their teenage years. In addition, a number of others mention that they "went underground" at this time, hiding their interest or disclosing it only to a few extremely close friends.

Approximately one of every three respondents is a woman, a figure that compares to one of every four adult American amateurs in the Young Entomologlsts Society. The extent to which either of these ratios accurately mirrors the numbers and sex distribution of amateur entomologists nationwide is unknown, of course. Nonetheless, it is still clear that the proportion of women among amateur entomologists is many times greater than the proportion of women among Ph.D. level professional entomologists in academia, which has been estimated at one in 70 (Wrensch 1987).

Despite increases in the past decade, the number of women in science is still small. A variety of processes operate to steer women away from the study of science as a basis for their life's work (White 1970, Berryman 1983). Attention to these processes is vital (Strickler 1987), but research protocols should be developed that separate interest in science from interest in a science career, because it appears probable that women are making decisions on these two issues independently.

How is an adult amateur's interest in entomology apt to relate to primary employment? Those in the sample show some striking similarities. More than 36% of the respondents are employed in sciences and mathematics. Often, they have a scientific background of a strength equal to that of many professional entomologists, but in a related field. Several have published in entomology or other biological areas. Another 17% are employed in education. They include public and private school teachers, park naturalists, nature center docents, and 4-H leaders. Whether or not their formal backgrounds include much entomology, their current knowledge of the subject is sufficient to form part of their day-to-day duties. Slightly less than 28% are in service occupations in which the biological sciences do not play a central role. In light of the major place that service occupations occupy in our society, this proportion may be less than one would expect by random assignment. Fewer than one in five respondents is employed in a "miscellaneous" nonassignable category. These respondents include a farmer, a retired woodworker, and a candy wholesaler. When one compares these occupations to the major employment patterns of paid entomologists, research in science, teaching college students, and employment in service occupations such as pest control, it is clear that the backgrounds and the current occupations of these respondents are far more similar to those of the professional entomologists than common stereotypes would lead one to believe.

What Motivates the Amateur?

The word "amateur" is derived from the Latin *amare* to love. In describing their avocation, amateurs uniformly use words like joy, excitement, wonder, delight, thrill, satisfaction, and fulfillment, and they share these feelings with others. A striking finding of the survey was the extent to which amateurs enthusiastically share entomology with children and the public at large. Only two respondents report no public expression of their entomological interests; the majority are involved in several amateur scientific activities. At the same time, even in response to a direct question, only three respondents report regular interaction with paid entomologists. One must conclude that the amateur entomologist is much more visible to the public than he or she is to the professional. Why?

First, the professional is also victim to stereotypes, in this case, that of an ivory tower recluse. Asked what they would like to share with professionals, almost half of the respondents specifically urge the professional to increase his or her interaction with nonscientists. The respondents are aware that as a nation, we are experiencing a significant decline in public support of and appreciation for science. The most recent large-scale study of public school students (Hueftle *et al.* 1983), a comprehensive survey of 18,000 children in three age groups, illustrates this dramatically in every attitude assessed.

Many science educators are pessimistic about the extent to which our schools are capable of reversing this trend. Elementary school students receive only minimal instruction in science (Conant 1974, Weiss 1978, Hurd 1978). Several surveys of middle and high school students, including one survey of 4,000 students in North Carolina (Simpson and Oliver 1985), suggest that the school science curriculum may even work against a student's chance of developing a continuing interest in science.

It is clear that in general, neither educators nor professional scientists are effectively communicating science's importance to this country's youth. It is the amateurs who are filling this vital role, and by and large they do not perceive professionals as being interested in sharing this work with them. Within professional circles, the extent and importance of the amateur's contribution to public appreciation of science and interest in its many manifestations is either unrecognized or ignored. Yet these are the professionals' public relations firm, the people who, out of personal joy and commitment, are cultivating the public support that indirectly, if not directly, keeps the professional employed.

Second, in their own interactions with professionals, most respondents report mixed receptions. Their perception is that many professionals are ill at ease with, if not cool or hostile toward adult amateurs. Perhaps, a few suggested, it is simply difficult to know how to deal with an adult who does work that is much like your own, but does not need to receive a paycheck to do it.

In a thought-provoking book entitled *Science Anxiety* physics professor Jeffry Mallow (1981) explores research on science attitudes and challenges scientists to resist the common temptation to treat science as if it were an elite subject for an elite population. Mallow proposes that many scientists, either consciously or unconsciously, send this message:

I am smarter than you. I was born smarter than you. There is
nothing you can do to comprehend science as well as I do. (p. 197)

Communicated to students, such arrogance is likely to produce
either science anxiety or downright avoidance of science. Based on
the responses to the survey, it appears that adult amateurs who
perceive this attitude in professionals are more apt to practice simple
avoidance, but otherwise their reactions are much the same as those
of the students.

How, then, might a professional scientist develop a better
relationship with amateurs? In a special Science Anxiety Clinic
established at Loyola University in Chicago, psychologists and
scientists found that a major improvement in interpersonal relations
resulted when a scientist changed the message to:

Here is science; it is something I have learned to do. You can learn
it too, because you are not very different from me. (Mallow 1981, p.
197)

This message is precisely the one that many amateurs naturally
send, and no doubt this is one reason they are often able to be such
effective communicators.

How Is an Amateur's Interest Nurtured?

A great many studies have firmly established the fact that interest
in science drops as children progress through school (Simpson and
Oliver 1985). The drop is particularly dramatic during the middle and
high school years, fueled by academic experiences and by peer
pressure (Hurd 1978, Talton and Simpson 1985). But what of those
who remain interested, but do not go on to become paid professionals? What strengths in their early experiences carry their interests
through the teen years into adulthood? At least among the amateur
entomologists, the answer is clear. Their continuing science interest
rarely flourishes because of contact with a teacher or professional
entomologist, not one of the respondents reported such a route.
Rather, it is nurtured by other amateur scientists, either within one's
own family or among one's acquaintances, ideally coupled with a
rich natural and educational environment.

A person's interest and enthusiastic support need not rest on any
detailed scientific background. John Miller, a political scientist at
Northern Illinois University in De Kalb, Illinois, has extensively
researched the subject of science attitudes among the adult population at large, concentrating on a trait which he calls attentiveness to
science (Miller and Prewitt 1979, Miller et al. 1980). He proposes that
this attentiveness in an adult's personality interacts with the
individual's science knowledge in a spiralling manner. The more

attentive one is, the more knowledge one absorbs from one's surroundings; the more knowledge one picks up, the more attentive one becomes.

Miller's observations dovetail nicely with the experiences that amateur entomologists report currently and recollect from childhood. Based on many amateurs' narrative responses, a contagious attitude of attentiveness on the part of those adults who have meaningful relationships with the child, rather than a specific family interest in entomology, appears to be the major factor predisposing a child to his or her own continuing interest in the subject. Even among adolescents, a review of 13 studies clearly indicates that adult interest and involvement continue to have a high correlation with a child's positive attitudes toward and interests in science (Kremer and Walberg 1981).

Is the Day of the Amateur Entomologist Gone?

Science in the nineteenth century was considered an appropriate avocation for cultured people. While employment in science may have been limited for some groups in society, popular knowledge of science was not. It is only in the past 40 years that science has taken on the aura of unexplained mystery available only to the initiated elite (Miller and Prewitt 1979, Mallow 1981).

Today, when so much of our country's future rests on the ability of its citizenry to make intelligent, informed decisions about science and technology, it is easy to become discouraged about the lack of widespread scientific literacy and support. This is particularly true if we look to the public schools to make progress in this area without any help from us. Despite lip service by the educational establishment, producing intelligent consumers of science has taken a back seat to an elitist search for potential producers of science (Simpson 1983). Led by academic scientists, much of the emphasis in science education has shifted toward preparing all young people for potential science careers (or at least to "weeding out" those "unsuited" for jobs in science and technology). This has resulted in science programs for adolescents that are irrelevant for large numbers of students at that age and maturity level (Hurd 1978). This situation exists despite the fact that although the number of salaried scientists our society can absorb has a limit, the number of enthusiastlc, supportive appreciators of science does not, and should not, have such a limit.

There are signs that we may be turning a corner. The past decade has seen serious attempts to bring science back to the people, and in general they have been well received. More and more articles about science are appearing in the general press. Television (particularly, but not exclusively, public television) has developed into an effective vehicle for the popularization of science. Many scientific societies now maintain public relations offices, and scientists doing pioneering work are routinely asked to hold press conferences. Three presidential addresses of the Lepidopterists Society have explicitly paid tribute to the amateur (Blanchard 1976, Miller 1986, Ferris 1986).

The public has responded positively to those individuals, amateurs and professionals alike, who see science as only one of a number of fascinating intellectual pursuits, rather than as the only interesting area of study or the exclusive province of a gifted few. Human enthusiasm is contagious. Grounded in the sort of feelings that the amateurs in this study describe, joy, excitement, and wonder, it can bridge the gap between professional and amateur, allowing both to welcome others into a fuller appreciation of science. I challenge you to join in sharing the fascination of discovery with others.

Acknowledgment

Grateful thanks go to the amateur emomologists across the country who shared their experiences. Without their help this commentary could not have been written. I also thank Julian P. Donahue and Richard D. Simpson for critical rewiew of the manuscript.

References Cited

Berryman, S. E. 1985. Who will do science? Rockefeller Foundation, New York.

Blanchard, A. 1976. Presidential address 1975 - to my fellow amateurs. J. Lepid. Soc. 30(1): 1-4.

Conant, E. 1974. What do teachers do all day? Sat. Rev. World 1: 55.

Evans, H. E. 1985. The pleasures of entomology: Portraits of insects and the people who study them. Smithsonian Institution Press, Washington, D. C.

Ferris, C. D. 1986. Presidential address. 1986: unexplored horizons - the role of the amateur lepidopterist. J. Lepid. Soc. 40(4): 247-254.

Hueftle, S. J., S. J. Rakow, and W. W. Welch. 1983. Image of science: a summary of results from the 1981-82 National Assessment in Science. University of Minnescta, Minneapolis.

Hurd, P. D. 1978. Final report of the National Science Foundation early adolescence panel meeting. National Science Foundation, Washington, D.C.

Kremer, B. K. and H. J. Walberg. 1981. A synthesis of social and psychological influences on science learning. Sci. Educ. 65(1): 11-23.

Mallow, J. V. 1981. Science anxiety: fear of science and how to overcome it. Van Nostrand Reinhold, New York.

Matthews, J. R. 1988. What makes an amateur tick? Young Entomol. Soc. Quart. 5(2): 9-19.

Miller, L. D. 1986. Presidential address, 1984: a tribute to the amateur. J. Lepid. Soc 40(1): 1-7.

Miller, J. D. and K. Prewitt. 1979. The measurement of the attitudes of the U.S. public toward organized science. National Science Foundation, Washington, D.C.

Miller, J., R. Suchner, and A. Voelker 1980. Citizenship in an age of science. Pergamon, Elmsford, N.Y.

Simpson, R. D. 1983. National attentiveness to science: an educational imperative for the eighties and beyond. Education 104(1): 7-11.

Simpson, R. D. and J. S. Oliver 1985. Attitudes toward science and achievement motivation profiles of male and female science students in grades six through ten. Sci. Educ. 69(4): 511-526.

Strickler, K. L. 1987. Women in entomology: increasing persistance in the pipeline. Bull. Entomol. Soc. Am. 33(1): 19-21.

Talton, E. L. and R. D. Simpson. 1985. Relationships between peer and individual attitudes toward science among adolescent students. Sci. Educ. 69(1): 19-24.

Weiss, I. R. 1978. Report of the 1977 national survey of science, mathematics, and social studies education. National Science Foundation Report SE 78-72. Government Printing Office, Wshington, D.C.

White, M. S. 1970. Psychological and social barriers to women in science. Science 170: 413.

Wrensch, D. L. 1987. Women in entomology: Deterministic road or random walk to professionalism? Bull. Entomol. Soc. Am. 33(1): 9-12.

Used with permission from: Bull. Entomol. Soc. Amer. 34(4): 157-161 (1988).

Mites or Acari

Edward C. Baker

Those minute arthropods, called mites, belong to the order Acari, which also includes the ticks. Spiders, daddy-long legs, scorpions, and similar arthropods are near relatives. Mites are very small, only about 1/200 of an inch on up to 1/2 inch in some species, but their relatives, the ticks, are much larger and better known. They are ectoparasites of domestic and wild animals, and occasionally man. The body of mites, unlike most other arthropods, have the head, thorax, and abdomen fused into a single body. Their mouthparts form a separate unit adapted for chewing or sucking.

Adult mites usually have four pairs of legs, as do the nymphal stages, but the larvae have only three pair of legs, as do insects, with which they can hardly be confused because of their distinctive shape. The life stages consist of larva, protonymph, deutonymph, and adult male and female. Many exceptions to the number of pairs of legs exist. Some males have three pairs and the female four. The plant gall mites belonging to the family Eriophyidae have only two pairs of legs in both sexes. Females of some tracheal mites have only a single pair of legs. Bodies may be round, oval, elongate, and even worm-like. Body surfaces may be smooth or spiculate, sparsely or densely covered with long or short setae. Colors vary from dirty gray, pearly white, brown, bright red, or multicolored. Reproduction is by male-female-egg development, or parthenogenesis without males.

Mites are found in practically all habitats. Their worldwide distribution, from the arctic to antarctic, makes them almost ubiquitous. Mites live on plants, on insects, on man, on animals, both domesticated and wild, or they may be found in brackish and fresh water, in hot springs, in houses, and in foodstuffs. As parasites skin follicles and eye lids of man and animals, nasal passages of humming birds, trachael systems of insects, and even bird feathers seem to be favorite habitats. Mites are the dominant arthropods in pastures, arable soil, and organic debris. Their associations with other animals include predation and parasitism. Many cause serious damage to livestock. They parasitize honey bee broods causing severe losses of bees and honey. Certain species, however, are important in the biological control of stored product and plant pests. Plant feeding mites of the spider mite and gall mite families include many pest species of economic importance. Their feeding can cause serious damage to agricultural crops, ornamental, and forest plants. Eriophyid mites feeding can cause severe plant abnormalities known as leaf galls, rusts, erineum, and bud blisters; they can also transmit plant virses, while others cause fruit distortion and inhibit plant growth.

Mites are associated with a wide variety of other animals. Beehives and bees support a large number of different species of mites. The pollen, honey and hive bottom debris including wax attract mites. Beetles, bumble bees and wasps are often found carrying mites on their bodies, some species in the abdominal pouches or acarinarium provided by the bees. Mites may also form intimate associations with animals, living on their bodies and obtaining food from gland secretions and skin debris. Small mammals carry mites in their fur; snakes and lizards under their scales. Birds have mites living in their quills, feeding on desquamated cells that give rise to skin irritation and desquamation.

Mites act as vectors of bacterial and virus diseases of their host and may spread the infection to man. Tsutsugamushi disease or scrub typhus in Japan and Asia is transmitted by trombiculid larvae. These larval mites feed normally on small rodents, which are the normal hosts of the rickesttsial pathogens. Rickettsial pox in man is also transmitted by a mite that normally lives on mice.

A number of mite species live in houses and stored foods, such as the typical cheese mite that show a preference for food of high fat and protein content. Others that feed on molds or fungi can cause serious infestation in damp dwellings, mattresses, and upholstered furniture; they are plentiful in house and mattress dust causing allergic reactions or rhinitis and house dust allergy.

Many more mite-plant and animals associations are known. The above are just a few examples.

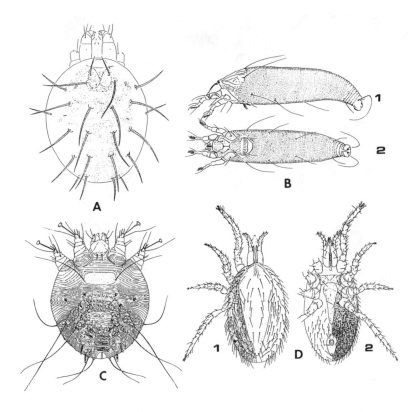

Figure 1. A. A plant feeding spider mite. B. An eriophyid or gall mite that forms galls on plants: (1) side or lateral view; (2) bottom or ventral view. C. A sarcoptes mite that feeds on man and animals. D. A rat mite that feeds on man and animals: (1) Upper or dorsal view; (2) bottom or ventral view.

Appendix I

American Registry
of Professional Entomologists
# of Entomological Society of America

The American Registry of Professional Entomologists of the Entomological Society of America is the primary registry of entomological expertise in North America. The Registry exists, by virtue of its charter granted by the Society, to benefit professional entomologists, and the public.

The mission of the Registry is to provide leadership for:

- The professional growth and well-being of entomologists throughout the Americas and other world communities.
- The development and publication to the public of a list of Registered Professional Entomologists.
- The development and implementation of an ongoing, certified in-service education program made available to members of the Registry and other interested scientists, technologists, or citizens.
- The publication and administration of a code of ethical conduct for Registered Professional Entomologists.
- The development and publication of a program of public education focusing upon the contributions that entomologists make to the well-being of all world communities.

Appendix II

Purposes

Today's society demands evidence of professional competence from those whose responsibilities affect the general well-being and economy as well as the quality of the environment. Entomologists are called upon to render a variety of services and certify the acceptability of procedures. New laws govern restrictions of choice and use of pesticides, and an increasing number of regulations limit the application of management procedures to qualified individuals. Local, State, and Federal government programs on pest management and regulatory procedure benefit from formal evidence of special competence. Thus, there is a clear and present need for a national registry to identify qualified individuals who provide technical service to the public in activities related to Entomology.

To meet this need, the Entomological Society of America has established a professional registry, identified as The American Registry of Professional Entomologists. The Registry has several chapters and associations comprised of registrants in localized areas. Those listed in the Registry have met high educational and practical experience standards and qualify for particular identification of special abilities. The board of registration is the 35-member Examining Board of the Registry.

Appendix III

The Entomological Society of America

The Entomological Society of America is a non-profit organization serving the scientific and professional needs of entomologists and individuals in related disciplines throughout the world. The Society has more than 9000 members and is the largest international association of entomologists. Numerous scientific publications are available for its members including four important journals — Annals of the Entomological Society of America, Journal of Economic Entomology, Journal of Medical Entomology, and Environmental Entomology - and two series of monographs - Miscellaneous Publications and the Thomas Say Foundation series. Annually, the Society holds national and regional conferences where members exchange scientific information, enhance their own professional knowledge and skills, conduct society business, and meet with colleagues from all corners of the world. These conferences are especially important for students and young professionals looking for employment and new contacts. The Society Placement Service also assists members in locating jobs for qualified personnel and keeping abreast of career opportunities in the field of entomology.

Appendix IV

American Mosquito Control Association

An international association of mosquito workers, entomologists, medical personnel, engineers, public health officials, military officers and personnel, and laymen who are charged with, or interested in the biology and control of mosquitoes and other vectors.

A non-profit, technical, scientific, and educational association, the purpose of which is to promote closer cooperation among those directly or indirectly concerned with, or interested in mosquito control and related work; to work for the highest standards of efficiency in such work; to encourage further research; to disseminate information about mosquitoes and their control; to work for understanding recognition and cooperation from public officials and from the public; to encourage the enactment of legislation providing for a sound, well balanced program of mosquito control work suited to local conditions wherever needed; to meet fairly and understandingly, and thus disarm opposition to mosquito control work from any source; to protect wildlife in every possible way from avoidable harm, and to encourage the use of control measures calculated to bring about the best practicable degree of adjustment where diverse interests are involved; to work for the highest degree of understanding cooperation with related organizations, to the end that the best interests of all may be most fully served; and to publish the Journal of the American Mosquito Control Association in the furtherance of these objectives.